Exploration Geophysics

Mamdouh R. Gadallah · Ray Fisher

Exploration Geophysics

 Springer

Mamdouh R. Gadallah
1120 Nantucket Drive
Houston, TX 77057
USA
mgadallah@comcast.net

Ray Fisher
14203 Townshire Drive
Houston, TX 77088
USA
rfisherosu@sbcglobal.net

ISBN: 978-3-540-85159-2   e-ISBN: 978-3-540-85160-8

DOI 10.1007/978-3-540-85160-8

Library of Congress Control Number: 2008934487

© Springer-Verlag Berlin Heidelberg 2009

Cover design: deblik, Berlin

Printed on acid-free paper

9 8 7 6 5 4 3 2 1

springer.com

Foreword

Today, we see that worldwide reserves are staying about the same, even increasing in some areas, partly because of the increased use of advanced technology in the exploration and development methods. Much of the credit for maintaining worldwide petroleum reserves must be credited to the 3-D seismic method. 3-D seismic surveys have resulted in the discovery of new fields, their development and enhancement of oil recovery projects. In addition, surface seismic surveys have been augmented by downhole surveys (VSP) that are used for borehole measurements of rock parameters such as density, acoustic velocity, and other parameters.

Another development of note has been the integration of historically separate personnel into teams of seismologists, geologists and engineers who are involved in all stages of petroleum exploration and exploitation. This has led to a need for all members of the team, their support staffs, and managers to better understand all of the technologies involved. The objectives of this text are to help satisfy this need for the non-professional members of these teams and those who support these teams in various ways.

This text will acquaint the people mentioned above with the fundamentals of the seismic techniques, their applications and limitations, with absolute minimal use of mathematics. The material is organized so that basic principles are followed by a flow of information paralleling that of applications. "Real-life" exercises are included to assist the understanding. *The text is written at a level that anyone can understand without difficulty.* At the end of each chapter you will find a list of key words that will help the reader to better understand the chapter by looking them up in the glossary at the end of the text. For those who are interested in more details, there are appendixes for some chapters that include more detailed information and a very complete bibliography of references for those who want to pursue the subject further.

Preface

It seems like digging the past but I still remember the day when the second edition of Dobrin's book "Introduction to Geophysical Prospecting" appeared. Even in those days, presenting this subject in a single volume was no easy task. Since then, our knowledge and capabilities for discovering oil and other natural resources has undergone a sea change. The credit mostly goes to a large number of individuals who used their highly specialized knowledge to analyze and solve a vast diversity of problems. Their contributions are well documented and continue to appear in technical books and professional journals. However, most of them are primarily useful for those engaged in research and development. Other professionals often find them too specialized or highly mathematical in nature. "Exploration Geophysics" by Gadallah and Fisher is a timely product which will fill a much needed gap.

The authors endeavor to present a simplified version of the science of exploration geophysics. In line with their professional background, they have primarily dealt with the various aspects of seismic prospecting. However, they cover almost everything related to this subject. After a short description of nonseismic methods, the reader is first introduced to an important but relatively less familiar subject of seeking permit for the acquisition of field data. This follows a detailed discussion on the acquisition and processing of data by using as little mathematics as possible. Much of the remaining book deals with the migration and interpretation of seismic data as well as the various tools needed to accomplish these tasks such as velocity analysis and the use of borehole information. In order to present a complete picture, the authors do not hesitate to touch upon the most recent developments such as cross-hole tomography and 4-D seismic.

The book provides a broad outline of seismic exploration without burdening the reader with nitty-gritty details. On the other hand, the door is kept open for further study by providing a comprehensive list of technical articles at the end of various chapters. At the other end of the spectrum, those quite new to the subject will find several lists of exercises valuable for self-learning. The book may also prove useful to those who work closely with geophysicists such as geologists, petroleum engineers as well as exploration managers.

Houston, TX   *Irshad Mufti*

Acknowledgement

We are grateful for all who so kindly allowed us to use some of their illustrations in our book. Specifically, we thank:

Society of Exploration Geophysicists
WesternGeco
American Association of Petroleum Geologists (AAPG)
Seismograph Services Corporation
CGG of America

We also wish to thank professors, friends, and colleagues who, through the years, have shared their knowledge and expertise with us. The contributions of these people made this book possible.

We also thank our wives, Jean Gadallah and Ileaine Fisher, for their patience, understanding, and encouragement.

Contents

List of Figures

Chapter 1
Introduction

The exact age of the earth is not known, but it is thought to be at least 4.5 billion years old Rocks and fossils (the remains of plants and animals preserved in the rocks) can be dated by measuring the decay rate of radioactive material that they contain. The number of radioactive particles given off by a substance during a certain time period provides a surprisingly accurate estimate of the age of the substance.

The geologic past is measured by means of a geologic time chart. Each interval of time has been given a name so that a particular time in the past can be referred to more easily. Periods in history are referred to in terms such as "the ice age" "the iron age" and "the atomic age". These time periods are measured in centuries or millennia at most. Intervals of geologic time, by contrast are measured in millions of years. For example, the dinosaurs became extinct about 70 million years ago. Another way to express it is "dinosaurs died at the end of the Cretaceous period."

Over the nearly 5 billion years of earth's (See Table 1.1) history mountains have risen, been eroded away and extreme environmental changes have occurred. For example:

- Palm tree fossils have been found near the north pole, indicating that a warm climate prevailed there in the geologic past
- Shark teeth have been found hundreds of miles from the nearest modern sea
- Many places that are high and dry today were once covered by seas. In fact many areas have been covered by seas, uplifted above sea level and submerged again multiple times. Such areas are now called basins

When rivers flow into a large body of water, suspended and dissolved sediments settle to the bottom. The coarsest sediments, such as sand, are deposited first and nearest to the river's mouth. Lighter sediments, such as mud and silt, are deposited farther out and in deeper water. Lime (calcium carbonate), produced by tiny life forms living in warm, shallow water, is deposited on the water bottom.

This deposition of sediments has occurred throughout geologic times that surface water has been present. Deposited sand is compacted and cemented to form sandstone. Lime hardens into limestone. These two sedimentary rock types are the rocks most important to petroleum accumulation and production.

Today, the oceans are teeming with life that ranges from giant whales and sharks to microscopic and near-microscopic size. As is the case now, the oceans were

M.R. Gadallah, R. Fisher, *Exploration Geophysics*,
DOI 10.1007/978-3-540-85160-8_1, © Springer-Verlag Berlin Heidelberg 2009

Era	Period	Epoch	
			today
	Quaternary	Holocene	
		Pleistocene	
Cenozoic			1.8 million years ago
	Tertiary	Pliocene Miocene Oligocene Eocene Paleocene	
			65 million years ago
Mesozoic	Cretaceous Jurassic Triassic		
			248 million years ago
Paleozoic	Permian Pennsylvanian Mississippian Devonian Silurian Ordovician Cambrain		
			543 million years ago
Precambrian			
			4.5 billion years ago

teeming with life millions of years ago. Many of these small life-forms, that were able to avoid being eaten, died and sunk to the ocean bottom where they were buried by mud and silt being deposited there. Immense quantities of organic material built within thick layers of mud and silt over millions of years.

Parts of the earth are constantly rising (being uplifted) or falling (subsiding) because of forces acting within the earth's upper layers. These are called tectonic forces. The uplifted parts cause the ocean shore to extend farther out. Sand transported in rivers gets deposited farther out in the ocean and covers the organic-rich layers of mud and silt. Later, the land subsides again and layers of mud and silt rich in organic material are deposited on top of the sand. The sequence of uplift, deposition, subsidence and more deposition has been repeated over and over again throughout the earth's history.

Since the organic material deposited on the ocean bottom is so quickly covered a process called anaerobic decay (without oxygen) occurs. The end products of this decay include the molecules of hydrogen and carbon (hydrocarbons) that make up oil and gas. Over a period of about one million years, the organic material is converted to petroleum. According to experts in the field, it takes 200^3 feet of dead organisms to make one cubic inch of oil.

Many people think that oil and gas are found in huge, cave-like caverns beneath the surface. This concept is completely wrong. In order to understand where oil and gas came from, how it is accumulated in place, and how to look for it, it is important to realize how very, very old the earth is and how many changes have taken place.

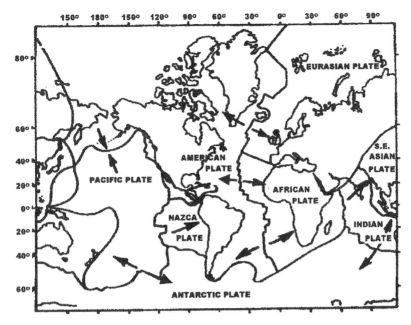

Fig. 1.1 Tectonic plates

Many theories to account for geologic activity have been proposed. The most satisfactory is one called *plate tectonic theory*. This was first proposed in 1967. The main idea is that the lithosphere (the crust and uppermost part of the mantle, averaging about 45 km thickness) is divided into large pieces, called plates that move relative to one another. The theory has been modified over the years, primarily by increasing the number of plates, now thought to number 28 (See Fig. 1.1).

New crust is formed at the crests of mid-ocean ridges and rises, resulting in what is called sea floor spreading. Old crust is destroyed by being plunged into the mantle beneath other plates along consuming plate boundaries. These plate boundaries are where there are deep oceanic trenches alongside island arcs or near mountain ranges along continental margins (See Fig. 1.2). Other plate boundaries are either extensional (plates are pulled apart as along the oceanic ridges) or transform (plates move horizontaally past one another, e.g. – along the San Andreas Fault of California).

Convection currents within the upper mantle provide the forces that cause plate motion. The asthenosphere, composed of a hot viscous liquid rock and is identified by a low seismic velocity, allows the plates to move over it. Spreading rates of the plates range from one to seven inches per year. Most earth scientists believe that at one time there was only one continent, named Pangea (See Fig. 1.3). Around 200 million years ago Pangea began breaking up because of plate motion. The continents we know today are composed of pieces of Pangea which have moved into their current positions.

Oil was first found in oil seeps where it accumulates on the surface of streams and lakes. Surface geology methods of petroleum were used for a while but it became

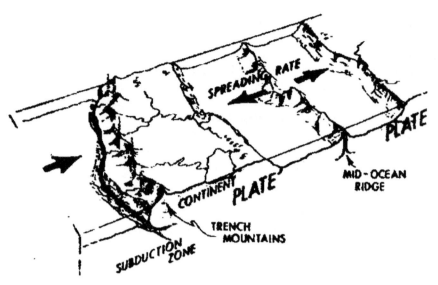

Fig. 1.2 Sea-floor spreading, the mechanism for tectonic plate motion

Fig. 1.3 Pangea

increasingly harder to find oil in these ways. More powerful and reliable techniques employing gravity, magnetic and seismic measurements have replaced the older methods, as well as the use of "water witches" to locate oil drilling sites.

Magnetic exploration for minerals, including, petroleum, is based on finding anomalous measurements of the earth's magnetic field. Similarly, measured anomalies in the earth's gravity can indicate the presence of subsurface geologic situations conducive to the accumulation of petroleum. More details of both methods and their applications to petroleum exploration are provided in Chap. 2.

The most reliable and most used technique of petroleum exploration is the seismic method. This involves recording "earthquake waves" produced artificially by explosives or some other energy source. Downward-traveling energy produces "echoes" at boundaries between rock layers. Determination of the times at which these "echoes" return to the surface supplemented by other information such as seismic propagation velocities allows an interpreter to develop a picture of the subsurface below the area investigated. More information about seismic waves and the seismic method is provided in Chaps. 2 and 3.

Chapter 2
Overview of Geophysical Techniques

Introduction

Various geophysical surveying methods have been and are used on land and off-shore. Each of these methods measures something that is related to subsurface rocks and their geologic configurations. Rocks and minerals in the earth vary in several ways. These include:

- *Density* – mass per unit volume. The *gravity method* detects lateral variations in density. Both lateral and vertical density variations are important in the *seismic method.*
- *Magnetic susceptibility* – the amount of magnetization in a substance exposed to a magnetic field. The *magnetic method* detects horizontal variations in susceptibility.
- *Propagation velocity* – the rate at which sound or seismic waves are transmitted in the earth. It is these variations, horizontal and vertical, that make the seismic method applicable to petroleum exploration.
- *Resistivity* and *induced polarization* – Resitivity is a measure of the ability to conduct electricity and induced polarization is frequency-dependent variation in resistivity. *Electrical methods* detect variations of these over a surface area
- *Self-potential* - ability to generate an electrical voltage. Electrical methods also measure this over a surface area.
- *Electromagnetic wave reflectivity and transmissivity* – reflection and transmission of electromagnetic radiation, such as radar, radio waves and infrared radiation, is the basis of *electromagnetic methods.*

The primary advantages of the gravity and magnetic methods are that they are faster and cheaper than the seismic method. However, they do not provide the detailed information about the subsurface that the seismic method, particularly seismic reflection, does. There may also be interpretational ambiguities present.

Electrical methods are well suited to tracking the subsurface water table and locating water-bearing sands Seismic methods can also be used for this purpose.

Electromagnetic methods are useful in detecting near surface features such as ancient rivers.

M.R. Gadallah, R. Fisher, *Exploration Geophysics,*
DOI 10.1007/978-3-540-85160-8_2, © Springer-Verlag Berlin Heidelberg 2009

There will be no further discussion of electrical or electromagnetic methods. The following paragraphs provide brief introductions to the gravity, magnetic, and seismic methods. The discussions of the gravity and magnetic methods included in this chapter serve to acquaint the reader with their general methods and applications. All subsequent chapters will deal with the seismic method.

The Gravity Method

A 70-kg man weighs less than 70 kg in Denver, Colorado and more than 70 kg pounds in Death Valley, California. This is because Denver is at a substantially higher elevation than sea level while Death Valley is below sea level. So, the farther from the center of the earth the less one weighs. What one weighs depends on the force of gravity at that spot and the force of gravity varies with elevation, rock densities, latitude, and topography. Mass, however, does not depend on gravity but is a fundamental quantity throughout the universe.

When a mass is suspended from a spring, the amount the spring stretches is proportional to the force of gravity. This force, F, is given by $F = mg$, where g is the *acceleration of gravity*. Since mass is a constant, variations in stretch of the spring can be used to determine variations in the acceleration of gravity, g.

Figure 2.1 illustrates the principle of gravity exploration. On the left the surface elevation is moderate but there is a thick sedimentary section overlaying the basement complex. At the center the surface elevation is near sea level and the subsurface has a sedimentary section of normal thickness and density overlaying an "average" basement complex. On the right the surface elevation is also moderate but there is a thin sedimentary section resulting in the basement complex being close to the surface.

The center part of Fig. 2.1 represents the "normal" earth situation and the suspended mass stretches the spring a "normal" amount here. On the left, the thick

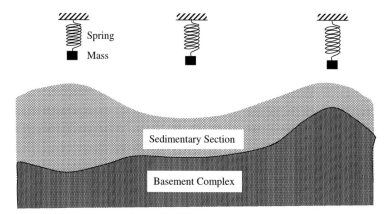

Fig. 2.1 The gravity method

sedimentary section has lower density than the basement rocks so the "pull" of the earth is reduced, resulting in the suspended mass stretching the spring less than the "normal" amount. The situation on the right is the opposite. The higher density basement rocks closer to the surface causes the "pull" of the earth to be greater, stretching the spring more than the "normal" amount.

An instrument called a *gravimeter* is used to measure g at "stations" spaced more or less evenly over the area being surveyed. Raw readings are corrected for elevation, latitude, and topography. The normal value of g is subtracted from the corrected readings, yielding *residual gravity*. The values of residual gravity are plotted at the respective station locations and contours of equal residual gravity are drawn. Closed contours represent *gravity anomalies* that can be used to infer subsurface geologic structures.

The value of g at sea level is about $980 \mathrm{cm/s^2}$. Since the variations of g are relatively small, $\mathrm{cm/s^2}$ is a bit large for measuring them. The unit used for measuring residual gravity is the *milligal* (*mgal*), or one-thousandth of a *gal*, where a gal is $1 \mathrm{cm/s^2}$. The gal is named for Galileo. Figure 2.2 is an example of a final gravity map. Contour interval is 1 mgal.

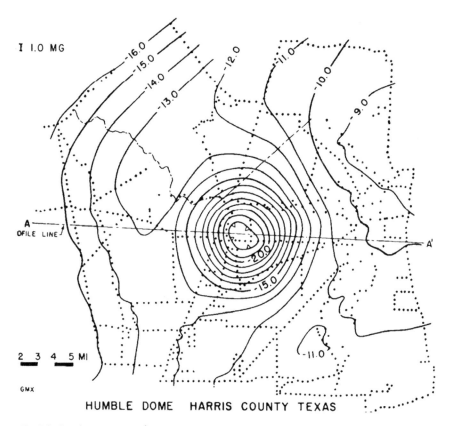

Fig. 2.2 Gravity map example

Interpretation of gravity data is done by comparing the shape and size of these anomalies to those caused by bodies of various geometrical shapes at different depths and differing densities.

The Magnetic Method

The earth's outer core is made of molten iron and nickel. Convection currents in the core result in motion of charged particles in a conductor, producing a magnetic field. The field behaves as though there is a north magnetic pole in the southern hemisphere and a south magnetic pole in the northern hemisphere. However, the magnetic poles of the earth are not coincident with the geographic poles of its axis. Currently, the north magnetic pole is located in the Canadian Northwest Territories northwest of Hudson Bay and the south magnetic pole is near the edge of the Antarctic continent. Note that the positions of the magnetic poles are not fixed but constantly change. The magnetic poles drift to the west at the rate of 19–24 km per year.

As a result of the shifting poles there is a change in the direction of the field, referred to as a secular variation. This is a periodic variation with a period of 960 years. In addition there are annual and diurnal, or daily, variations.

A magnetic field can be described by magnetic lines of force that are invisible. These lines can be thought of as flowing out of the south magnetic pole and into the north magnetic pole. A compass needle aligns itself along the magnetic line of force that passes through it. If the compass needle were free to move vertically as well as horizontally, it would point vertically downward at the north magnetic pole, vertically upward at the south magnetic pole and at intermediate angles away from the magnetic poles. Figure 2.3 illustrates the earth's magnetic field. A compass needle aligns itself along the line of force passing through it.

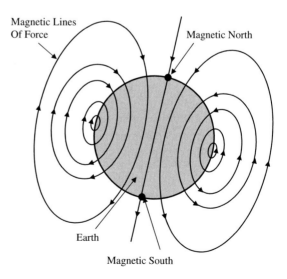

Fig. 2.3 Earth's magnetic field

In addition to these known variations in the magnetic field, local variations occur where the basement complex is close to the surface and where concentrations of ferromagnetic minerals exist. Thus, the primary applications of the magnetic method are in mapping the basement and locating ferromagnetic ore deposits.

The instruments used to measure the earth's magnetic field are called magnetometers. What is actually being measured is the *intensity* or *field strength* of the earth's field. This is measured in *Tesla (T)*. Since the objective of the magnetic method is to detect relatively small differences from the theoretical value of magnetic intensity, these are measured in *NanoTesla (nT)* or *gammas* (γ). $(1 \ nT = 10^{-9} \ T = 1 \ \gamma.)$

Today, most magnetic surveys are made from airplanes. While flying over a predetermined path (usually, a set of parallel flight lines), the magnetic field is continuously recorded. The raw magnetometer readings must be corrected for diurnal variations and other known causes of magnetic intensity variations. The residual field is determined by subtracting the theoretical values for the area of survey from the corrected magnetometer readings. The residuals are plotted on a map and contours of equal gammas are drawn. See Fig. 2.4.

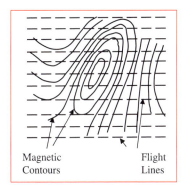

Magnetic   Flight
Contours   Lines

Fig. 2.4 Magnetic map

Closed contours indicate magnetic *anomalies* caused by local ferromagnetic bodies or anomalous depths to the basement. Interpretation is similar to that for gravity except the bodies of various geometrical shapes at different depths differ in magnetic susceptibilities rather than densities.

The Seismic Method

The seismic method is rather simple in concept. An energy source (dynamite in the early days) is used to produce seismic waves (similar to sound) that travel through the earth to detectors of motion, on land, or pressure, at sea. The detectors convert the motion or pressure variations to electricity that is recorded by electronic instruments.

L. Palmiere developed the first 'seismograph' in 1855. A seismograph is an in-strument used to detect and record earthquakes. This device was able to pick up and record the vibrations of the earth that occur during an earthquake. However, it wasn't until 1921 that this technology was used to help locate underground oil formations.

There are two paths between source and receiver of particular interest – reflection and refraction. In Fig. 2.5 layers 1 and 2 differ in rock type, in the rate at which seis-mic waves travel (*acoustic or seismic velocity*), and *density* (mass per unit volume). When the seismic waves encounter the boundary between layers 1 and 2 some of the energy is reflected back to the surface in layer 1 and some is transmitted into layer 2. If the seismic velocity of layer 2 is faster than in layer 1, there will be an angle at which the transmitted seismic wave is bent or refracted to travel along the boundary between layers, as shown in Fig. 2.5. These two path types are the bases of seismic *reflection* and *refraction* surveys.

Figure 2.6 illustrates seismic reflection operations. Instead of a single detector as in Fig. 2.5, 24 detectors are laid out on the surface. Seismic energy travels down-ward with some being reflected at the boundary between layers 1 and 2 back to the detectors. (Note: actual operations involve recording many reflections from many subsurface reflectors from many more detectors than are shown here.)

Reflection seismic data are displayed as *seismic records* consisting of several *seismic traces*. A seismic trace, often presented as a "wiggly line", represents the response of a single seismic detector (or connected group of detectors) to the earth's movement caused by the arrival of seismic energy. Figure 2.7 illustrates a simulated seismic reflection record that was developed from Fig. 2.6. (This is called a *shot record* because all traces represent energy from a single source or shot.) Traces are ordered by *offset* or distance from the source.

Similar "wiggles" can be followed from trace-to-trace starting at about 0.165 s on trace 1 and ending at about 0.78 s on trace 24. This *event* is the first break refraction. It is refraction from the base of the shallow near surface layer that is too thin to adequately show in Fig. 2.6. Note that a straight line can be drawn through this event.

A second event is shown in Fig. 2.7. This event, the reflection from the boundary between layers 1 and 2, starts at about 1.90 s on trace 1 and ends at about 1.99 s on trace 24. Note that it is not straight but curved.

A seismic reflection survey generates a large number of shot records that cover the area under study. Modern methods call for recording reflections such that there is a common midpoint between sources and detectors on many different shot records.

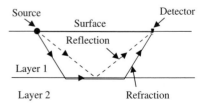

Fig. 2.5 Reflection and refraction

Fig. 2.6 Seismic reflection method

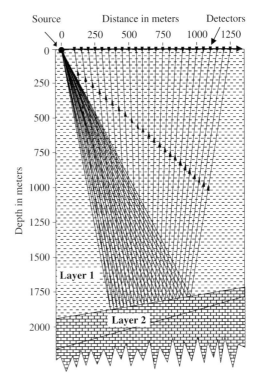

Fig. 2.7 A simulated seismic reflection record, based on Fig. 2.6

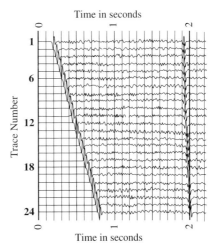

In seismic data processing the traces that share these common midpoints are collected together as *common midpoint* or *CMP records*. The assumption is that these traces record from the same subsurface reflection points and are combined, or *stacked*, into a single trace, called a *CMP trace*. Other processes are applied to the data to enhance the signal, minimize noise, and improve interpretability.

Fig. 2.8 A seismic section

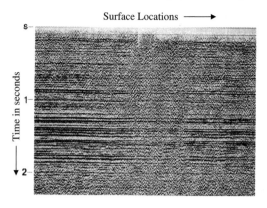

When processing is complete, all the CMP traces are displayed side by side comprising a *seismic section*. The section is an image of the subsurface, that can be used to plan drilling and development programs. The section in Fig. 2.8 shows many rock beds and a potentially hydrocarbon-bearing structure.

The reflection method has been the most successful seismic method for identifying subsurface geologic conditions favorable to the accumulation of oil and gas. The greater part of this book discusses and explains this method.

Figure 2.9 illustrates the seismic refraction method. Here, seismic waves travel faster in layer 2 than in layer 1, i.e. – *seismic velocity* is higher in layer 2 than in layer 1. The seismic waves that arrive at the layer boundary at the *critical angle* are bent or refracted along the boundary. At the receiver end, seismic waves are refracted upward at the same angle. Additional refractions may occur at deeper boundaries, if the seismic velocities below the boundaries are faster than those above the boundaries.

Figure 2.10 is a simulated seismic refraction record based on Fig. 2.9. Again two events are apparent. The first is the refraction from the boundary between layer 1 and 2. The second is the direct arrival from the source.

Less processing is applied to refraction data than reflection data. The main interest is in being able to pick the arrival time of refraction events. These times are

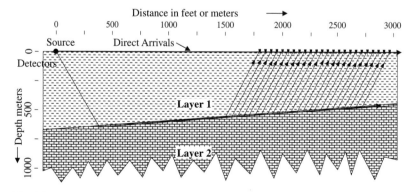

Fig. 2.9 Seismic refraction method

Fig. 2.10 A simulated
seismic refraction record,
based on Fig. 2.9

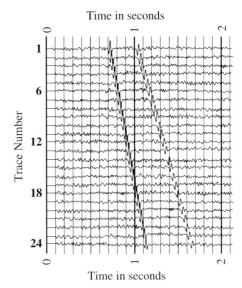

plotted against offsets (distances between source and receivers) in what are called T-X plots. Analysis and interpretation of these plots may allow determination of subsurface layer thicknesses and velocities.

The refraction method can supply data that allow interpreters to identify rock units, if the acoustic velocities are known. The refraction method can also be used to detail structure of certain deep, high-velocity sediments, where reflection data are not of sufficient quality.

Summary and Discussion

This chapter provides a brief review of the geophysical methods used in petroleum exploration and development. Chapter 3 gives the basic theory and principles upon which the seismic method is based. Chapter 4 covers seismic refraction surveys in somewhat greater depth. The rest of the book covers various aspects of seismic reflection methods.

Gravity and magnetic methods can be used for reconnaissance surveys to delineate areas of interest. They should be conducted before (or in conjunction with) the seismic method.

Today high-resolution 3-D seismic data are used to delineate petroleum reservoirs before drilling commences, determine optimum locations for initial drilling, select sites for development wells, and to monitor reservoirs throughout their various production cycles.

The seismic industry continues to develop ever more sophisticated methods. These are needed to allow discovery of petroleum deposits to replace depleted reserves. The more subtle nature of the reservoirs to be discovered, require more accurate information so that the fine details of a reservoir can be studied. These advanced methods are also needed to optimize petroleum production from known reservoirs.

There are many sources of data and information for the geologist and geophysicist in exploration for hydrocarbons. This includes a variety of measurements, commonly referred to as logs, obtained along the boreholes. However, this raw data alone would be useless without methodical processing and interpretation. Much like putting together a puzzle, the geophysicist uses sources of data available to create a model, or educated guess, as to the structure of rocks under the ground. Some techniques, including seismic exploration, allow the construction of a hand or computer generated visual interpretation of the subsurface. Other sources of data, such as that obtained from core samples or logging, are taken by the geologist when determining the subsurface geological structures. It must be remembered, however, that despite the amazing evolution of technology and exploration methods the only way of being sure that a petroleum or natural gas reservoir exists is to drill. The result of the improvement in technology and procedures is that exploration geologists and geophysicists can make better assessments of drilling locations.

Chapter 3
Seismic Fundamentals

Basic Concepts

It is necessary to introduce some basic concepts before discussing seismic methods. That is the purpose of this chapter

Seismic Waves

The principle of sound propagation, while it can be very complex, is familiar. Consider a pebble dropped in still water. When it hits the water's surface, ripples can be seen propagating away from the center in circular patterns that get progressively larger in diameter. A close look shows that the water particles do not physically travel away from where the pebble was dropped. Instead they displace adjacent particles vertically then return to their original positions. The energy imparted to the water by the pebble's dropping is transmitted along the surface of the water by continuous and progressive displacement of adjacent water particles. A similar process can be visualized in the vertical plane, indicating that wave propagation is a three-dimensional phenomenon.

Types of Seismic Waves

Sound propagates through the air as changes in air pressure. Air molecules are alternately compressed (compressions) and pulled apart (rarefactions) as sound travels through the air. This phenomenon is often called a sound wave but also as a compressional wave, a longitudinal wave, or a P-wave. The latter designation will be used most often in this book.

Figure 3.1 illustrates P-wave propagation. Darkened areas indicate compressions. The positions of the compression at times t_1 through $6t_1$ are shown from top to bottom. Note that the pulse propagates a distance d_p over a time of $6t_1 - t_1 = 5t_1$. The distance traveled divided by the time taken is the propagation velocity, symbolized V_p for P-waves.

M.R. Gadallah, R. Fisher, *Exploration Geophysics*,
DOI 10.1007/978-3-540-85160-8_3, © Springer-Verlag Berlin Heidelberg 2009

Fig. 3.1 Propagation of a
P-wave pulse

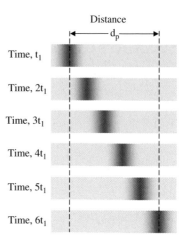

P-waves can propagate in solids, liquids, and gasses. There is another kind of seismic wave that propagates only in solids. This is called a *shear wave* or an *S-wave*. The latter term is preferred in this book. Motion induced by the S-wave is perpendicular to the direction of propagation, i.e. – up and down or side-to-side.

Figure 3.2 illustrates propagation of an S-wave pulse. Note that the S-wave propagates a distance d_s in the time $5t_1$. The S-wave velocity, designated as V_s, is $d_s/5t_1$. Since d_s is less than d_p, it can be seen that $V_s, < V_p$. That is, S-waves propagate more slowly than P-waves.

Surface waves are another kind of seismic waves that exist at the boundary of the propagating medium. The *Rayleigh wave* is one kind of a surface wave. It exhibits a retrograde elliptical particle motion. Figure 3.3 shows motion of a particle over one period as a Rayleigh waves propagates from left to right. The Rayleigh wave is often recorded on seismic records taken on land. It is then usually called *ground roll*. Love waves are similar surface wave in which the particle motion is similar to S-waves. However, Love wave motion is only parallel to the surface.

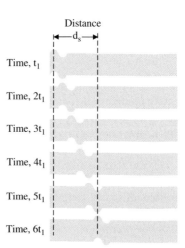

Fig. 3.2 Propagation of an
S-wave pulse

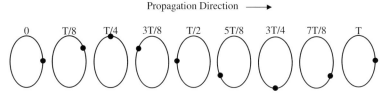

Fig. 3.3 Rayleigh wave particle motion

Seismic Wave Propagation

In comparing seismic wave propagation to the wave generated around a pebble thrown in the water, replace the pebble with a device such as an explosive or vibrator that introduces energy into the ground. This energy initially propagates as expanding spherical shells through the earth. A photograph of the traveling wave motion taken at a particular time would show a connected set of disturbances a certain distance from the source. This leading edge of the energy is called a *wave front.* Many investigations of seismic wave propagation in three dimensions are best done by the use of wavefronts.

Beginning at the source and connecting equivalent points on successive wave fronts by perpendicular lines, gives the directional description of wave propagation. The connecting lines form a *ray,* which is a simple representation of a three-dimensional phenomenon. Remember, when we use a ray diagram we are referring to the wave propagation in that particular direction; that is, the wave fronts are perpendicular to the ray at all points (see Fig. 3.4).

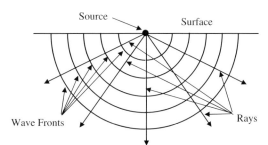

Fig. 3.4 Wave fronts and rays

Reflection and Refraction

As a first departure from the simplest earth model, consider a layered earth. What happens when an incident compressional wave strikes a boundary between two media with different velocities of wave propagation and/or different densities? Answer: Part of the energy is reflected from the boundary and the rest is transmitted into the next layer. The sum of the reflected and transmitted amplitudes is equal to the incident amplitude.

The relative sizes of the transmitted and reflected amplitudes depend on the contrast in *acoustic impedances* of the rocks on each side of the interface. While it is difficult to precisely relate acoustic impedance to actual rock properties, usually the harder the rocks the larger the acoustic impedance at their interface.

The acoustic impedance of a rock is determined by multiplying its density by its P-wave velocity, i.e., V. Acoustic impedance is generally designated as Z.

Consider a P-wave of amplitude A_0 that is normally incident on an interface between two layers having seismic impedances (product of velocity and density) of Z_1 and Z_2 (See Fig. 3.5). The result is a transmitted ray of amplitude A_2 that travels on through the interface in the same direction as the incident ray, and a reflected ray of amplitude A_1 that returns to the source along the path of the incident ray.

The reflection coefficient R is the ratio of the amplitude A_1 of the reflected ray to the amplitude A_o of the incident ray,

$$R = \frac{A_1}{A_0} \tag{3.1}$$

The magnitude and polarity of the reflection coefficient depends on the difference between seismic impedances of layers 1 and 2, Z_1 and Z_2. Large differences $(Z_2 - Z_1)$ in seismic impedances results in relatively large reflection coefficients. If the seismic impedance of layer 1 is larger than that of layer 2, the reflection coefficient is negative and the polarity of the reflected wave is reversed. Some Typical values of reflection coefficients for near-surface reflectors and some good subsurface reflectors are shown below:

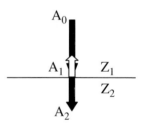

Fig. 3.5 Normal reflection
and transmission

It can be seen in Table 3.1 that a soft, muddy ocean bottom reflects only about one-third of the incident energy, while a hard bottom reflects about two thirds of the energy.

The transmission coefficient is the ratio of the amplitude transmitted to the incident amplitude:

$$T = \frac{A_2}{A_0} = 1 - R \tag{3.2}$$

When a P-ray strikes an interface at an angle other than 90°, reflected and transmitted P-rays are generated as in the case of normal incidence. In such cases, however, some of the incident P-wave energy is converted into reflected and transmitted S-waves (see Fig. 3.6). The resulting S-waves, called SV waves, are polarized in the vertical plane. The Zoeppritz' equations are a relatively complex set of equations that allow calculation of the amplitudes of the two reflected and the two transmitted

Table 3.1 Typical reflection coefficients

Near-Surface Reflectors:	
Soft ocean bottom (sand/shale)	0.33
Hard ocean bottom	0.67
Base of weathered layer	0.63
Good Subsurface Reflectors	
Sand/shale versus limestone at 4,000 ft	0.21
Shale versus basement at 12,000 ft.	0.29
Gas sand versus shale at 4,000 ft.	0.23
Gas sand versus shale at 12,000 ft.	0.125

Fig. 3.6 Reflection and re-fraction of an incident P-wave. $V_{P2} > V_{S2} > V_{P1} > V_{S1}$

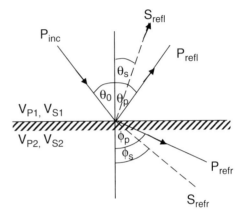

waves as functions of the angle of incidence. The equations require P- and S-wave velocities (V_{P2}, V_{S2}, V_{P1}, and V_{S1} in Fig. 3.6) plus densities on both sides of the boundary. The S-waves that are called converted rays contain information that can help identify fractured zones in reservoir rocks but this book will discuss compressional waves only.

Snell's Law

This relationship was originally developed in the study of optics. It does, however, apply equally well to seismic waves. Its major application is to determine angles of reflection and refraction from the incidence of seismic waves on layer boundaries at angles other than 90°.

Snell's law of reflection states that the angle at which a ray is reflected is equal to the angle of incidence. Both the angle of incidence and the angle of reflection are measured from the normal to the boundary between two layers having different seismic impedances.

The portion of incident energy that is transmitted through the boundary and into the second layer with changed direction of propagation is called *a refracted ray.* The direction of the refracted ray depends upon the ratio of the velocities in the two layers. If the velocity in layer 2 is faster than that of layer 1, the refracted ray is bent toward the horizontal. If the velocity in layer 2 is slower than that of layer 1, the refracted ray is bent toward the vertical.

Table 3.2 Snell's law relationships

Velocity relationship	Angle relationship
$V_{P2} > V_{S2} > V_{P1} > V_{S1}$	$\phi_p > \phi_s > \theta_0 > \theta_s$
$V_{P2} > V_{P1} > V_{S2} > V_{S1}$	$\phi_p > \theta_0 > \phi_s > \theta_s$
$V_{P1} > V_{P2} > V_{S1} > V_{S2}$	$\theta_0 > \phi_p > \theta_s > \phi_s$
$V_{P1} > V_{S1} > V_{P2} > V_{S2}$	$\theta_0 > \theta_s > \phi_p > \phi_s$

Figure 3.6 illustrates the more general condition for reflection and refraction. In this case both P- and S-wave velocities on each side of the interface are specified because reflected P- and S-waves and refracted P- and S-waves are generated from the incident P-wave. The two angles of reflection depend on the ratios V_{P1}/V_{P1} and V_{S1}/V_{P1}. The ratio of 1 for the reflected P-wave is a restatement of the angle of reflection equaling the angle of refraction for the P-wave. Since S-wave velocity is always slower than P-wave velocity the reflected S-wave always reflects at an angle less than that of the P-wave. The two angles of refraction depend on the ratios V_{P2}/V_{P1} and V_{S2}/V_{P1}. The relationships between angles of reflection and refraction with velocity ratio are not simple ones but depend upon the trigonometric function sine of the angles.

In Fig. 3.6 the relationships among the various velocities are: $V_{P2} > V_{S2} > V_{P1} > V_{S1}$. As a result the angles of refraction for both P- and S-waves are greater than the angle of incidence. There are, however, three other possible relationships. They are shown in Table 3.2, along with the corresponding relationships among angles of refraction. (Angles of reflection are not affected).

Critical Angle and Head Waves

From Table 3.2 it can be seen that when the P-wave velocity is higher in the underlying layer, the refracted P-ray is "bent" toward the boundary. As the angle of incidence increases the refracted P-ray will be bent to where it is just below and along the boundary, which means that the angle of refraction is 90°. The particular angle of incidence at which this occurs is known as the *critical angle,* usually designated θ_c. The sine of the critical angle is equal to the ratio of velocities across the boundary or interface.

This wave, known as *a head wave,* passes up obliquely through the upper layer toward the surface, as shown in Fig. 3.7.

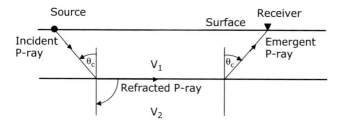

Fig. 3.7 Critical refraction/head wave

Fermat's Principle

A seismic pulse that travels in a medium follows a connected path between the source and a particular receiver. However, according to Fermat's principle there is the possibility of multiple travel paths. That means there may be more than one primary reflection event. The *buried focus* ("bow tie") effect is a classic example of Fermat's principle. On the left of Fig. 3.8 is the representation of a deep syncline and ray paths to and from seven coincident receivers and sources. There is only one path for rays numbered 1 and 7. There are two paths for rays 2, 3, 5 and 6. There are three paths for ray 4. On the right the arrival times are plotted vertically below the source/receivers. Note the crossing images and apparent anticline that results. This feature could be mistaken for a real anticline and a well that results in a dry hole.

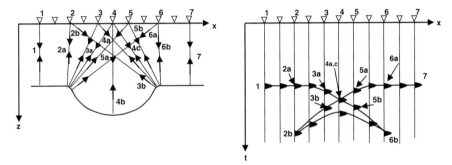

Fig. 3.8 On the *left* is a sketch of a deep syncline (buried focus) and reflection ray paths. On the *right* is its appearance on a seismic section (bowtie effect)

Huygens's Principle

This principle states that "Every point on an advancing wavefront is a new source of spherical waves". The position of the wave front at a later instant can be found by constructing a surface tangent to all secondary wavelets. See Fig. 3.9. Huygen's Principle provides a mechanism by which a propagating seismic pulse loses energy with depth.

Attenuation of Seismic Waves

As seismic waves propagate over greater and greater distances the amplitudes become smaller and smaller. That is, seismic waves are attenuated with the distance traveled. On a seismic record, this appears as attenuation with record time.

Even in a perfect medium, seismic waves are attenuated with distance. Consider the analogy of a balloon. Initially, the balloon is opaque. As the balloon becomes

Fig. 3.9 Huygen's principle

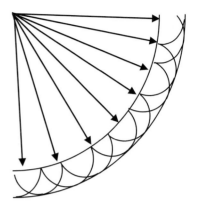

more fully inflated its color becomes lighter until it is almost transparent. (See Fig. 3.10). This is because the balloon gets thinner and thinner as it gets bigger and bigger. There is just as much material in the balloon as before it was inflated, it's just thinner.

As seismic waves propagate away from the source the wavefront that describes the wave's advance becomes larger and larger. The energy gets spread over an ever-larger surface area. As a result, energy per unit area becomes smaller. Seismic amplitudes are proportional to the square root of energy per unit area so amplitudes get smaller even at a greater rate than the decrease in energy per unit area. This type of amplitude attenuation is called *spherical spreading* or *geometrical spreading*.

Another reason that seismic amplitudes get smaller is that rocks are not perfect conductors of seismic energy. Rocks are made up of individual particles or crystals. As a result, some of the energy becomes scattered. It does not all go in the main direction of propagation. In addition, because seismic wave propagation involves motion of particles, there is some "rubbing" of rock particles against one another. This results in some seismic energy being converted to heat. The higher the frequency of the seismic waves the greater the heat loss, and scattering, that occurs. This means that seismic wavelets become lower in frequency and longer in duration the farther they travel and hence, the later they arrive at the seismic detectors. This type of amplitude attenuation is called *inelastic attenuation*. Figure 3.11 illustrates the effect of both geometrical spreading and inelastic attenuation.

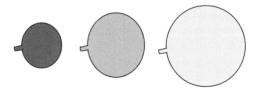

Fig. 3.10 Effect of balloon inflation

Fig. 3.11 Change in reflection amplitude with record
time

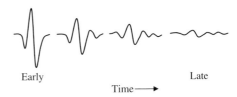

Early   Late

Time ⟶

Propagation Model for Exploration Seismology

Exploration seismology was developed to explore sedimentary basins that have gentle dip and layered structure with horizontal continuity over a large area. Simple
models that include these essential features and propagating seismic pulses in these
models enhance the understanding and interpreting of seismic records and sections.

The models adopted here assume that the seismic energy propagates along paths
involving multiple receivers and multiple sources. The following propagation models will make it clear that the redundancy in sources and receivers allow estimation
of needed velocity information.

Figure 3.12 is the simplest model considered. It consists of a single layer overlying a semi-infinite medium with the layer boundary being flat and horizontal. The
thickness of the layer is Z and its propagation velocity has a constant value of V. This
model can be used to calculate time required for energy to travel from the source to
the receiver via reflection from the base of the layer.

There is an energy source at S and 12 receivers laid out at equal intervals, or
offsets, from the source. Reflection raypaths are straight lines down to the base of
the layer and straight lines up to the receivers. Reflection points are midway between
source and receiver on the reflector. Reflection times are simply the total lengths of
these pairs of lines divided by the velocity, V.

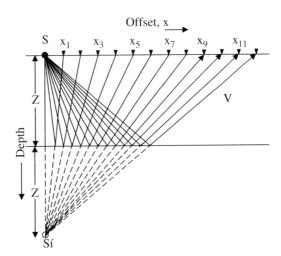

Fig. 3.12 Simple earth model

Constructing the image point of the source at S', which is at a depth Z below the boundary or a distance 2Z from the surface perpendicular to the surface, allows drawing of the dashed lines shown in Fig. 3.12. The lines from S' to x_1 through x_{12} are the same lengths as the two segments of the ray paths from S to x_1 through x_{12}. Figure 3.13(a) shows the lengths of the raypaths S'x_1 through S'x_{12} designated by d_1 through d_{12}. Note that while the lines increase in length with increasing offset, the rate of increase is not linear. If a curve is drawn connecting the ends of the lines representing reflection path length, it is found to be a curved line called a hyperbola.

As previously noted, for the constant velocity layer of Fig. 3.12, reflection times are given by dividing total path length, d, by velocity V. Thus, time for the reflection recorded at x_1 is $T_1 = d_1/V$. The time for the reflection recorded at x_2 is $T_2 = d_2/V$. Times T_3 through T_{12} are calculated by dividing d_3 through d_{12} by V. Figure 3.13(b) plots the reflection times corresponding to the reflection raypaths of Fig. 3.12. Trace number corresponds to number of the receiver from which data were recorded.

The zero-offset time, T_0, is defined as the time required for a vertical reflection from the source to the base of the layer and back. Expressed as an equation, $T_0 = 2Z/V$. The reflection times T_1 through T_{12} are all greater than T_0. Thus, these times can be expressed as

$$T_j = T_0 + \Delta T_j, \quad j = 1, 2, \ldots, 12.$$

The quantity ΔT_j is called normal moveout or NMO and it depends on both the offset and velocity. It also has a hyperbolic shape. One of the important seismic data

(a)   (b)

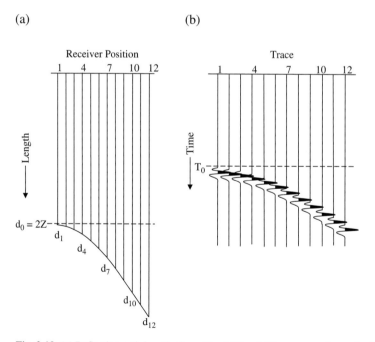

Fig. 3.13 (a) Reflection path lengths from Fig. 3.12 and (b) corresponding reflection times

processes is the correction for NMO, allowing true reflection times to be determined. Since the real earth is much more complex than the simple model of Fig. 3.12, and seismic velocities are not generally known beforehand, considerable effort is expended to extract velocity information from the data.

Summary and Discussion

Seismic waves propagate in three dimensions and following a seismic pulse through the earth is a difficult task. To better understand this propagation process, the pulses are followed through greatly simplified earth models.

Seismic waves occur as compressional waves, or P-waves, shear waves, or S-waves, and Rayleigh waves. P-waves are usually of greatest interest. S-waves can be used to obtain more detailed or special information about the subsurface. Rayleigh waves may be recorded on land seismic records as ground roll, an undesirable event and, hence referred to as "noise". P-waves propagate in solids, liquids and gasses. S-waves propagate only in solids. P-waves always have higher propagation velocities than S-waves, in the same medium.

Seismic energy that is input to the ground using an energy source such as an explosive (e.g.; dynamite) or a vibratory source (Vibroseis$^{®}$*) energy propagates outward from the source in expanding spheres through the earth. Surfaces of these spheres are called wavefronts

Seismic rays indicate the paths that seismic waves take between two or more points in a medium. They are always perpendicular to the wavefronts. It should be remembered that when a ray diagram is presented, it implies wavefronts that are perpendicular to the ray at all points.

Parts of the earth of interest in petroleum exploration are made up of many layers, or strata, that have different geological and geophysical properties. Of particular interest are propagation velocity and density. The product of these two is called acoustic impedance. When a P-wave is incident on a boundary between these layers some energy is reflected and some is transmitted. Reflection and transmission coefficients are ratios of reflected and transmitted amplitudes.

The seismic method adapts to the theory of optics to study the propagation of the seismic energy in the earth. Snell's law of reflection and refraction is fundamental to understanding the seismic energy propagation. Huygens's principle provides a view of seismic energy propagation and attenuation. Fermat's principle introduces the possibility of multiple travel paths between source and receiver that may give rise to more than one primary reflection event.

Exercises

1. Name and describe three types of seismic waves described in this chapter.
2. Define the following terms:

a. Acoustic impedance
b. Snell's law of refraction
c. Critical angle

3. Table 3.3 lists densities and velocities of three layers. What can you infer about the magnitudes and polarities of reflection coefficient 1 (for the interface between layers 1 and 2) and reflection coefficient 2 (for the interface between layers 2 and 3)?

Table 3.3 Densities and velocities for earth model

Layer	Density (gm/cm^3)	Velocity (m/s)
1	2.2	1500
2	2.9	3000
3	2.6	2500

4. Consider two reflectors, or interfaces between two layers. In the first case, the velocity of the upper layer is 2.5 km/s and the velocity of the lower layer is 5.0 km/s. In the second case, the velocity of the upper layer is 3.25 km/s and the velocity of the lower layer is 4.75 km/s if a ray travels downward through the top layer at an angle of incidence of 20° in each case, which will result in a larger angle of refraction.
5. Below a flat, horizontal surface is a layer of 1500 m thickness that has a constant velocity of 2500 m/s. Twelve detectors are placed at 100 m intervals from the source. Table 3.4 lists total path length to each detector. Determine T_0 and NMO (ΔT) for traces corresponding to each detector. List answers in ms. (1 ms = 1000 s.)

Table 3.4 Reflection path lengths

Detector	Offset (m)	Reflection path lengths (m)	ΔT (ms)
1	100	3001.7	
2	200	3006.7	
3	300	3015.0	
4	400	3026.5	
5	500	3041.4	
6	600	3059.4	
7	700	3080.6	
8	800	3104.8	
9	900	3132.1	
10	1000	3162.3	
11	1100	3195.3	
12	1200	3236.1	

Bibliography

Bath, M.*Introduction to Seismology.*Basel-Stuttgart: Birkhauser Verlag, (1973).

Birch, F. "Compressibility, Elastic Constants." S. P. Clark, ed., *Geological Society of America Memoir 97* (1966):97–173.

Dix, C. H. "Seismic Velocities from Surface Measurements." *Geophysics* 20 (1955):68–86.

Ewing, M., W. Jardetzky, and F. Press. *Elastic Waves in Layered Media* New York: McGraw-Hill, (1957).

Faust, L. Y. "Seismic Velocity as a Function of Depth and Geological Time." *Geophysics* 16 (1951):192–196.

Gardner, G. H. F., L. W. Gardner, and A. R. Gregory. "Formation Velocity and Density – the Diagnostic Basis for Stratigraphic Traps." *Geophysics 39* (1974):770–780.

Koefoed, O. "Reflection and Transmission Coefficients for Plane Longitudinal Incident Waves." *Geophysics Prospect 10* (1962):304–351.

Muskat, M., and M. W. Meres. "Reflection and Transmission Coefficients for Plane Waves in Elastic Media." *Geophysics 5* (1940):115–148.

Sharma, P. V. *Geophysical Methods in Geology.* Amsterdam: Elsevier, (1976).

Sheriff, R E. "Addendum to Glossary of Terms used in Geophysical Exploration." *Geophysics 34* (1969):255–270.

Telford, W. M., L. P. Geldart, R E. Sheriff, and D. A. Keys. *Applied Geophysics.*Cambridge: Cambridge University Press, (1976).

Trorey, A. W. "A Simple Theory for Seismic Diffractions." *Geophysics 35* (1970).

Chapter 4
Data Acquisition

Introduction

A successful seismic data acquisition program requires careful and detailed planning before fieldwork begins. Such planning should include the following steps:

- Select and describe primary and secondary targets
  The primary target must be described in terms of its location, geologic type, depth, areal extent, and expected dips, particularly the maximum dip. Similar information about the secondary target should be specified. This target should be shallower than the primary one. It is mostly used as a reference and control surface.
- Estimate potential production and profits
  Obviously, anticipated profits must exceed costs of acquiring, processing, and interpreting the seismic data as well as drilling and other exploitation costs. If the estimated production is not expected to provide such profits, there is no point in going further.
- Budget acquisition costs
  A total budget, i.e. for acquisition and processing, must be determined first. Generally more funds are allocated for acquisition than processing. Acquisition may constitute up to 80% of the total budget.
- Specify and document program objectives and priorities
  A contractor is usually chosen to carry out the acquisition program. Often this is done through competitive bidding. In order to make an intelligent bid, the contractor must know what is expected. Priorities are necessary to provide for unanticipated situations that preclude realizing all objectives within the allotted budget and time.
- Establish data quality standards
  Quality standards must be selected such that the desired objectives can be met, consistent with budget and time constraints.
- Set reasonable schedules and deadlines
  The contractor must know when the program is to start, how long it is to take, and intermediate progression requirements. The client needs established production requirements to evaluate the contractor's performance.

M.R. Gadallah, R. Fisher, *Exploration Geophysics*,
DOI 10.1007/978-3-540-85160-8_4, © Springer-Verlag Berlin Heidelberg 2009

- Locate desired lines of survey on maps
  Having defined and described the targets and determined objectives, desired surface positions and spacing between them that satisfy objectives can be drawn on a map of the area. It must be understood that modifications in the desired locations may be necessary because of permitting, access, and other problems found by inspection of the area.
- Select specific methods and equipment to be used
  Choices depend on environment (land or marine, terrain, surface conditions, etc.), acquisition parameters required to meet program objectives, personnel and equipment availability, tightness of schedule, and cost.

Permitting

Practically every bit of the earth's surface belongs to a person or an entity such as a corporation, government or religious group. Before seismic operations can begin, it is necessary to gain permission to work from these property owners. The person who does this is often called a "Permit Man" or "Permit Agent".

Maps developed in the initial planning stage indicate where the "properties" are. The first thing the Permit Agent must do is to determine who owns the "properties". For marine surveys various departments and/or agencies of nations or political subdivisions thereof (states, provinces, etc.) are the owners. In the case of land operations in the United States, taxing agents in the county court houses are the best source of property information. A complication can arise in many parts of the U.S. when a piece of land may have one owner of the surface and another who owns its subsurface "mineral' rights". A further complication is that the person occupying the land may be leasing it from the owner. Outside the U.S. most governments own all "mineral rights" but the surface land may be owned privately.

The Permit Agent must:

- Determine who all these owners are,
- Gain permission for seismic work from them, and
- Communicate to the field crew any restrictions imposed by the owners.

  Above all, the permitting work must be done expeditiously so work can begin.

Acquisition Requirements

Elements of a seismic reflection data acquisition system include the following:

1. **Surveying/navigation system -** Precise locations of source and receiver positions must be known.
2. **Energy sources -** Seismic waves having appropriate amplitudes and frequency spectra must be generated.

3. **Receivers** - Seismic waves must be detected and converted into electrical signals.
4. **Cables** -Signals output from the receivers must be transmitted to the recording system with minimum attenuation and distortion.
5. **Recording system**- Signals transmitted via the cables must be recorded in a form that provides easy retrieval while preserving as much as possible of the information contained in the original signal.

Surveying and Navigation

Desired lines of survey are established in the planning stage. In the case of land operations, the surveyor must determine the feasibility of positioning these lines in the desired locations and recommend modifications, if needed. Once the line positions, lengths, etc. are determined the surveyor must locate these with regard to a *control point* or a known position on the earth's surface. In some more remote parts of the earth, it may be necessary to establish a control point. This is usually done by means of GPS (Global Positioning System).

GPS is a satellite-based positioning system that currently uses 27 satellites in orbit around the earth. What makes GPS so valuable in seismic work is that it can be used in all-weather conditions, it has very good accuracy over long distances, can be used 24 h a day just about anywhere, is very reliable, and is often much faster than the conventional surveying techniques.

The surveyor must determine the position and elevation of every source and receiver point in the survey with the required degree of accuracy. This is usually done within an x-y coordinate system, the origin of which is precisely located with respect to the selected or established control point.

The surveyor must also produce a variety of maps. A final map shows the positions of all source and receiver points. Maps must be provided to the source and recording crews that show fences, streams, ponds, structures, etc. Such maps should also show areas to be avoided because of hazards or lack permission for entry. Maps showing how to get to points across fences, streams, etc., including notes about gates and landowner restrictions must also be provided.

In marine work location of the energy source array and seismic detectors is done simultaneously with recording operations. The vessel location is directly determined with sources and receivers being determined relative to the vessel. Accurate positioning and steering of the vessel is required to obtain data where it is needed. Accurate positioning and steering of the vessel is also required to avoid numerous hazards (surface facilities, buoys reefs shoals, international boundaries) that are frequent.

Marine navigation relies on the observation of radio waves to determine the vessel position relative to precisely known reference positions known as "base stations". Both surface-based and satellite-based radio positioning systems are used. Surface-based systems use fixed base stations that are located on the surface of the earth near the prospect site. Satellite-based systems use orbiting satellites as the base stations, e.g. GPS.

Radio positioning systems are based on measurements of radio signal transit times or phase that can by converted to equivalent distances by scaling based on the propagation velocity of radio waves. Phase measurements have a cyclic ambiguity of an integer number of wavelengths.

Energy Sources

Desirable characteristics of seismic sources include:

- Signal – High amplitude, broad frequency bandwidth produced.
- Safety – Hazards in use, storage and maintenance can be managed without excessive precautions.
- Cost – Total cost of equipment (acquisition, operation and maintenance) and supplies must be considered.
- Operation – Relatively simple, efficient and fast operation generally preferred
- Environment – Minimal physical and biological damage to the surroundings produced by the source.

There are two basic types of energy sources, Impulsive and vibratory. Table 4.1 summarizes these.

When explosives are used they are most often loaded at the bottom of a drilled hole (or holes). This requires one or more drills mounted on trucks. In most areas the drills use a drilling fluid (mud) to cool the drills. This requires a truck to bring water to the drills so that the fluid can be made. Two-person crews are needed to operate the drills.

The charge is usually dynamite or ammonium nitrate fertilizer mixed with diesel fuel. The size of the charge depends on depth and shot medium. The preferred technique is to drill through the low-velocity zone (weathering). Principal advantages of this technique are that time through the low-velocity zone can be measured directly (via an uphole geophone). Consequently having only one pass through the low-velocity zone reduces signal attenuation and minimizes the generation of surface waves.

Sometimes the time and cost of drilling deep holes is just too much. In such cases, a large number of shallow shot holes are drilled. These holes are drilled in

Table 4.1 Energy source types

Source	Land	Marine	Comments
IMPULSIVE			
Explosives -			
Dynamite	√		Usually shot in drilled holes on
Ammonium Nitrate	√		land but rarely used as marine source, today
Geoflex/Primacord	√		Shot very near the surface
Airgun		√	Most popular marine source
VIBRATORY Vibroseis	√		Most popular land source

a geometrical pattern or array. This procedure enhances the signal and attenuates surface waves at the source.

Those who drill holes generally load them. A firing cap with a wire lead is inserted in the charge. The charge is lowered to the bottom of the hole and pushed down with a loading pole to secure it in place. The drilling fluid, mentioned above, generally fills the holes. Thus, in loading the holes care must be taken to avoid having the charge float to the top of the hole. After loading, the holes are plugged and covered until they are ready to be shot.

The recording crew follows after the drilling crew in land operations. In fact, it may be a number of days after the holes are drilled and loaded before the recording crew is ready to shoot the charge in the hole(s). The recording truck is positioned such that many shots can be fired before it has to be moved.

A person called a shooter goes to the locations of the holes (shot points) and communicates with the instrument operator. The shooter connects the cap which leads to a blaster and tells the instrument operator he is ready. When the instrument operator is ready, the recording instruments are started and a radio signal is transmitted to the blaster. See Fig. 4.1.

This starts a sequence of events that causes the charge to explode. A considerable part of the energy produced by the explosion often results in permanent deformation in the form of a cavity and cracks in the medium around the shot and may blow out material (e.g. drilling mud, rocks) from the hole. However, a significant part of the energy is transmitted as seismic waves in approximately spherical wavefronts radiating outward in all directions. See Fig. 4.2.

A seismic detector (uphole geophone) is placed near the top of the hole. Some of this energy reaches the uphole geophone via a minimum time path. This provides a direct measure of the time from the explosive to the surface. This is very valuable information used in seismic data processing.

Geoflex is an explosive cord plowed into the ground about 18″ deep. It is a small charge but is very efficient. Geoflex detonates at the rate of 21,000 ft/s. It is cheap and fast because no holes are drilled required but a soft surface is required. It also attenuates horizontal noise such as ground roll. See Fig. 4.3.

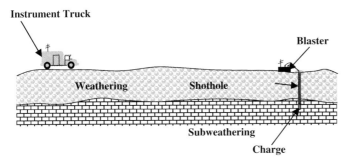

Fig. 4.1 Explosive technique

Fig. 4.2 Explosive source
operation

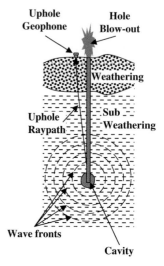

Fig. 4.3 Geoflex operation

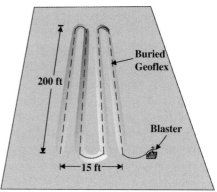

The principal advantage of explosives is that high energy and a broadband signal is produced. As mentioned earlier, another advantage is that a direct measure of time through low - velocity zone can be obtained when the explosives are shot in drilled holes.

Disadvantages are that much energy may be lost in blow - out of the hole and permanent deformation of material around the charge. Moreover, high amplitude horizontal noise is usually produced when explosives are shot at, above or just below the surface. Drilling trucks, auxiliary equipment, and supplies may be expensive. Personnel costs of drilling may be high, particularly when drilling in difficult areas where slow production may increase need for more drilling units. Strict safety regulations are imposed and tight security is required in the use and storage of explosives. Harmful effect of explosions on marine life all but eliminates its use as a marine source. There are many government regulations on use of explosives that must be followed.

Fig. 4.4 Vibrator pad during
sweep

A vibrator is a vehicle-mounted energy source that uses hydraulic energy to produce a signal that is usually several seconds long. The base plates, or "pads" are
lowered to the ground and trucks are jacked-up to place the weight of trucks on
the base plates, providing a reactive mass. The vibrator actuator converts hydraulic
energy into mechanical energy input to the base plate. See Fig. 4.4.

When vibrators are used to generate seismic energy, two to four (sometimes
more) vibrator trucks are positioned at source points within source array (patch).
An encoded swept-frequency signal (pilot sweep) is transmitted from the instrument
truck to the vibrator trucks. Figure 4.5 is an example of a pilot sweep.

All the vibrators send their sweep signals into the ground and the instruments
begin recording simultaneously. Recording continues for length of the sweep plus
the "listening" time. The vibrator trucks then release jacks, raise the plates and move
on to the next positions in the patch. The procedure just described is repeated at
these positions and further positions, as required to sweep at all source positions in
the patch.

In Fig. 4.5 the pilot sweep is seven seconds long. If the final record length is to
be 5 s long, the raw vibrator records must be 12 s long.

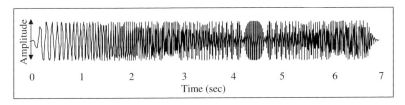

Fig. 4.5 Vibroseis pilot sweep example

Vibrators, as with most surface energy sources tend to produce large amplitude ground roll. A vibrator patch is an array (to be discussed later in this chapter). When all records obtained at a single shotpoint are summed together the array effect results in attenuation of this source-generated noise. This process of summing individual records is called vertical stack. In addition to attenuating source-generated noise, vertical stack also increases the signal strength relative to random noise.

The sweep is recorded twice – once, when is transmitted to the vibrator control units and, second, after passing through the same filters as the recorded data. The later version is called the "filtered sweep". The stacked record is crosscorrelated with the filtered sweep, producing a single output record, the length of which is equal to the listen time.

Figure 4.6 illustrates the crosscorrelation process. The vibrator sweep is shown at the top. Below it is the recorded raw trace with overlapping reflections of the vibrator sweep. It is difficult, if not impossible, to identify reflection signals on this trace. The crosscorrelation process compresses the long reflected sweep reflections into much smaller reflection wavelets. Shown below the raw trace is the zero-phase correlated trace. This is the direct output of crosscorrelation. In some cases this is undesirable and the trace is converted from zero-phase (symmetrical wavelets) into minimum-phase wavelets. Minimum phase wavelets are those that have the maximum amount of energy as close to the start of the wavelet as possible (amplitudes are highest at the front of the wavelet).

In summary:

- Vibrators allow the selection of signals' frequency content, which is usually a desirable thing. Available frequencies range from 5 Hz to 511 Hz.
- Sweep lengths can be up to 31 s.
- A wide variety of sweep types are available besides the linear sweep. Sweeps may be up (increasing frequency) or down (decreasing frequency). The sweep of Fig. 4.5 is a linear upsweep.
- Cosine taper is applied at each end of sweep to limit the side-lobe amplitudes.
- Vibrators cannot be used in marshy and mountainous areas or in jungles but can be used about anywhere else, including towns and cities,
- Very hard surfaces tend to distort vibrator signals.

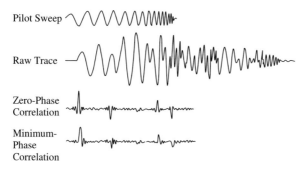

Fig. 4.6 Raw and correlated vibrator traces

Desirable marine sources:

- Generate a powerful pulse
- Fire rapidly (in approximately 10 s intervals)
- Operate simply, consistently, and dependably
- Maintain a constant depth and produce minimum drag when towed
- Do not injure marine life
- Have minimal repeated expansions and contractions of gas volume

Airguns are the usual choice because they are about the only marine source that meets all of the above requirements.

Figure 4.7 is a cut-away view of a type of airgun, usually called a *sleeve gun*, currently in use. Compressed air from a compressor on the back deck of the vessel enters through the air intake. The sleeve is down over the exhaust ports, initially. The firing chamber and the chamber are filled to the required pressure (2000 psi) via the fill passage and chamber fill orfice. At the proper time the sleeve moves upward releasing compressed air into the water via the exhaust ports, forming a bubble around the airgun.

A single airgun, however, does not produce adequate energy or a satisfactory pulse. The bubble from an airgun expands outward in all directions until it reaches its point of maximum expansion. However, the air pressure in the bubble is now less than that of the surrounding water. So, the bubble contracts to point until air pressure is again greater than that of the surrounding water. As a result, a second smaller expansion of the air bubble occurs followed by a second smaller contraction. Successive expansions and contractions continue until all energy is dissipated. Because of this *bubble effect,* the signal waveform from a single gun is very long, not the desired short impulsive waveform. See Fig. 4.8.

It is the introduction of compressed air at pressure substantially above water pressure that produces the signal. Increasing pressure above 2000 psi has not proven effective; so only two ways of increasing the airgun signal levels are available:

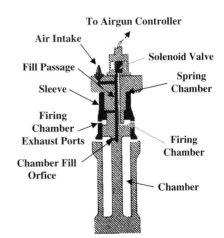

Fig. 4.7 Airgun components

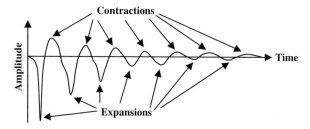

Fig. 4.8 The bubble effect

1. Increase Airgun volume. Since air bubble pressure is proportional to the cube root of gun volume, doubling the volume only increases pressure by 26%. Doubling the pressure requires increasing volume by a factor of 8.
2. Increase the number of guns. If two airguns of the same volume are placed close together, the pressure and signal amplitude are doubled. Increasing Airgun signal amplitudes is a desirable thing but does not solve the bubble problem nor does it produce a desired signal wavelet. Solution of this problem requires combining many airguns of different volumes separated at optimum distances. Figure 4.9 illustrates the methodology.

Figure 4.9 shows signatures (signal waveforms) for five different volumes with airgun volume decreasing from top to bottom. Note that the larger volume guns produce lower frequency signals. The smaller guns reach maximum expansion at an earlier time than the larger guns so delays are built in to allow all guns to reach maximum expansion simultaneously. The bottom of Fig. 4.9 shows the array signature resulting from an airgun array using many guns of these volumes. Note that the later bubble effects of the different volume guns occur out of phase from one another resulting in the combined array signature.

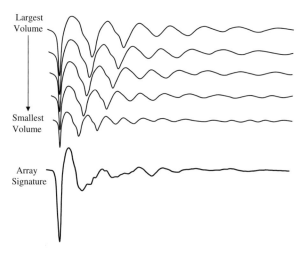

Fig. 4.9 Individual airgun and combined airgun array signatures

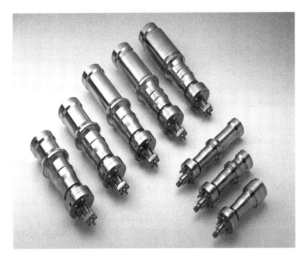

Fig. 4.10 A family of sleeve guns

The airguns used today are called sleeve guns. Figure 4.10 is a photograph showing sleeve guns of various sizes.

Summarizing, airguns:

- Produce signals by the release of compressed air into water
- Are most widely used marine seismic energy source because they are relatively cheap, are safer than explosives, and are not harmful to marine life
- Can be recharged very quickly
- Have frequency content that depends on depth below the surface, pressure of compressed air and volume of the airgun chamber
- Generate bubble by inertia overshoot
- Minimize harmful effect of bubbles and increase signal amplitudes when used in Airgun arrays

Seismic Receivers

Seismic energy is a form of mechanical energy. Seismic receivers convert this mechanical energy into electrical energy. Different devices to perform this conversion must be used, depending on the environment and physical quantity to be measured. In land operations the quantity being measured is ground motion. In marine operations it is pressure. Table 4.2 lists the different types of seismic receivers.

Figure 4.11 is a cutaway diagram of a geophone. The case holds working parts of the geophone and usually has a planting spike to hold the geophone to the ground so that it faithfully follows the motion of the ground. A spring arrangement assures that the mass and wire coil remain stationary with respect to the earth as a whole. Surrounding the coil, and fixed to the case is a permanent magnet. When the ground

Table 4.2 Seismic receivers

Type	Name(s)

Motion-Sensitive:
  (a) Velocitymeter – output is Proportional to
     ground velocity. (mv/in/s)
  (b) Accelerometer– output proportional to
     ground acceleration. (mv/in/s^2)
Pressure-Sensitive:
  Pressure Transducer (mv/µBar)
	Geophone or Seismometer (Jug, Seis, Seisphone)
	Note: Seismograph used in earthquake seismology.
	Accelerometer
	
	Hydrophone (water phone, crystal phone, pressure phone)

Fig. 4.11 Geophone components

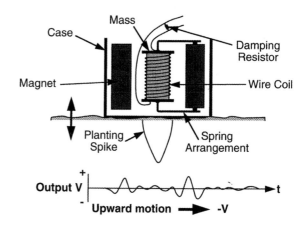

moves, the coil moves through the magnet's field producing a voltage that is proportional to relative motion between the mass and the case. This is the quantity to be measured. Note that it is the ground particle velocity that is being measured, not amplitude of ground motion.

If a mechanical system consisting of a mass suspended by a spring is set into motion, it will vibrate at a particular frequency that depends only on the size of the mass and the "stiffness" of the spring. This is called the *resonant or natural frequency*. The mass and spring of the geophone constitute such a system, so one of the geophone parameters is its resonant frequency, measured in Hertz (Hz).

Another geophone parameter is its *sensitivity*. This simply means that a unit ground velocity produces the voltage. The usual unit is mv/ips.

The last parameter is damping. Figure 4.11 shows a damping resistor connected across the two output wires. This causes some of the output current to be fed back to the coil, reducing the output voltage but also modifying geophone response in a desirable way as shown in Figure 4.12.

Without a damping resistor, the geophone has a large response at the resonant frequency. With proper damping, the response is quite flat over frequencies of interest. Obviously, a geophone must have a resonant frequency lower than the lowest frequency in the signal.

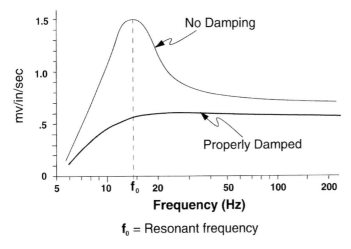

Fig. 4.12 Geophone responses

A geophone must be insensitive to temperature variations because they are used in many different locations and differing weather conditions. Similarly they must be waterproof and dust proof to be useful in all environments. Geophones must be reliable, light in weight, and easily handled, as they will be used many times and moved frequently. Since large numbers of geophones are used, they must be inexpensive. A high output is very desirable for ease in recording.

Hydrophones are used in the marine environment. Hydrophones employ piezo-electric crystals in the form of disks as their active elements. Piezoelectricity is the property that some materials have of producing an electric current when a stress is applied to it. Thus, when pressure is applied to the face of a piezoelectric disk, a current proportional to the pressure is produced. Unfortunately, a jerk on the streamer can also produce a current. To avoid having this happen hydrophones are made with pairs of piezoelectric disks, wired so that outputs from pressure variations add while those from acceleration of the streamer subtract. See Fig. 4.13.

Outputs of hydrophones are quite small so an array of several hydrophones is used. The hydrophones are spaced very closely together and the array effect is to increase signal, not attenuate noise.

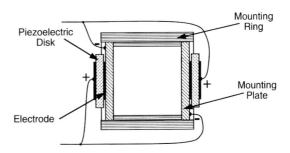

Fig. 4.13 Acceleration-canceling hydrophone

Seismic Arrays

An array is a group of two or more elements (sources or receivers) arranged in a geometrical pattern. The pattern may be one or two-dimensional (linear or areal). The function of arrays is to do spatial filtering. An array's response depends upon wavelength or wavenumber of seismic energy produced or received. Larger and more complex arrays use receivers rather than sources because of cost difference.

Signal amplitudes are generally increased (maximum signal attenuation of 3 dB) over frequency range of interest but horizontal, source-generated noise is attenuated by 20 dB or more. Signal-to-random noise ratio improvement is $\sqrt{N}$, where N is the number of elements in the array.

Figure 4.14 shows a few array types. Areal arrays should be used only when noise appears to come from directions other than along the line of receivers. In 3-D operations care must be exercised to avoid having the array attenuate the signal.

Figure 4.15 shows how an array attenuates horizontal noise but not the signal. Shown is a simple, six-element linear array. Noise produced by the source generally propagates horizontally at a rather slow velocity. Thus, there is a time delay between noise arrivals at successive geophones in the array. Outputs of each geophone are electrically summed, producing the outputs indicated by Σ. Signal arrives at the surface nearly vertically and at nearly the same time. Thus the summed signal output is nearly six times that of a single geophone.

Array performance is a shown in Fig. 4.15 only if the surface is relatively flat. If, however, there is considerable elevation variation across the array the result is not so good. Figure 4.16 shows an array with considerable elevation differences and compares its signal response with that without elevation differences. Signal no longer adds in phase, so arrays should not be used in such situations. It is better to bunch the geophones at the array center.

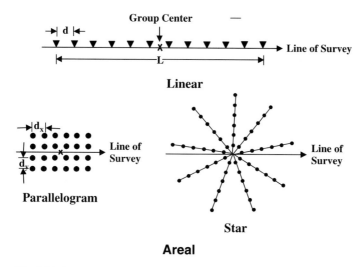

Fig. 4.14 Array types

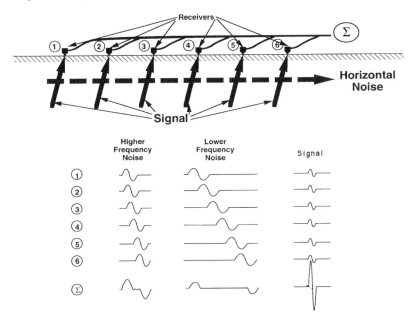

Fig. 4.15 Effect of arrays on signal and noise

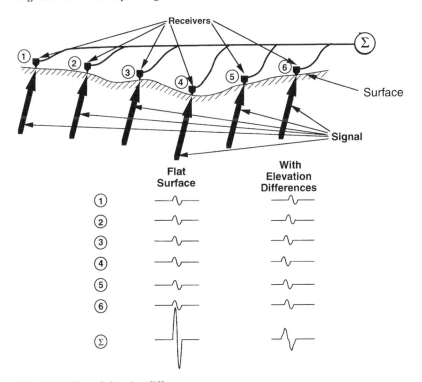

Fig. 4.16 Effect of elevation differences on array response

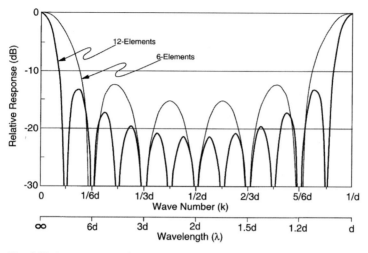

Fig. 4.17 Array responses for simple linear arrays

Figure 4.17 shows responses of simple linear arrays of 6 and 12 elements. Increasing the number of elements increases attenuation overall and sharpens the response nodes.

Some practical points to consider in design and use of arrays are:

- Detectors are made in strings of six, so the number used must be a multiple of six
- A finite number of strings are available, usually only enough for 4–6 per group
- There is a finite length of wire between detectors so there is a maximum spacing between detectors that can be used
- Elevation differences among detectors `may exist. Use of arrays may not be advisable
- Strings connections should be simple to avoid errors in connection. There is always a tendency to hurry so that high production is obtained.
- Two-dimensional arrays cannot always be laid out because of terrain or obstacles
- Layout is not very accurate since everything is "eyeballed in".
- Detectors may have unequal outputs resulting in poorer than expected responses.

These points lead to the conclusion that design of elaborate arrays and concern with details in layout, etc. represent wasted time and effort!

Recording Instruments

The purpose of recording instruments is to provide an uncontaminated, precise, permanent record of data detected by receivers in the spread so that data can be studied and analyzed at a later date. As shown in Fig. 4.18, amplitudes and the times at which the signal arrives are the data required. Unfortunately, the simple method of Fig. 4.18 will not get the job done.

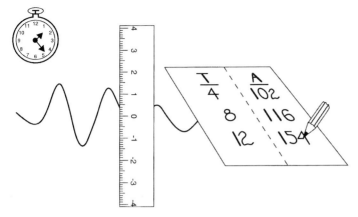

Fig. 4.18 Instrument function

Seismic recording systems used today have two distinct parts: ground or inwater systems, and truck or onboard systems. Ground (land) and inwater (Marine and OBC) systems modules are located near the receiver groups.

Figure 4.19 shows a configuration of a ground system as used in land 3-D operations. The line cables are sectionalized cables that connect to the "boxes" at each end. Receiver groups are connected to the line cables. The number of group connectors per cable depends on the number of channels handled by the boxes. All of the line cables are connected to a module called a Line Interface Unit in Fig. 4.19 It collects and transmits all inputs to the truck mounted system. .

Figure 4.20 shows the configuration of a *streamer*. A marine inwater system usually consists of several streamers. Streamers are held on cable reels on the stern

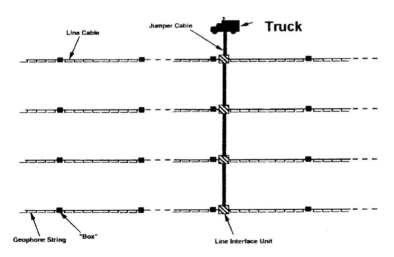

Fig. 4.19 Land ground system configuration

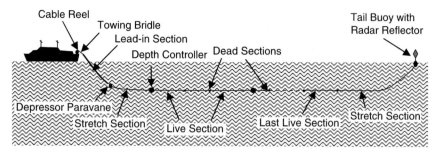

Fig. 4.20 Streamer configuration

of the vessel when not deployed. Electrical connection between the streamer electronics and the onboard system are made at these reels. The *towing bridle* and the *lead-in section* provide the physical connection between the vessel and the streamer proper.

Streamers also use sectionalized cables but there is more than one kind. Live sections have hydrophone groups built into them. Stretch sections are places at the front and back of the streamer to absorb shocks caused by tail buoy jerks, for example. Dead sections are used as spacers. In addition, depth controllers that monitor streamer depth and "birds" that are used to maintain and adjust streamer depth are attached to the streamer.

The equivalent of the boxes in land ground systems are also present in marine inwater systems. They are functionally, and perhaps electronically, identical to their land counterparts. They may be connected between cable sections or integrated in the live sections.

Streamers must have neutral buoyancy to maintain proper depth. Historically, this has been accomplished by introducing a fluid within the cables. Some streamer sections now employ solid material for this purpose.

A tail buoy with a radar reflector is placed at the end of each streamer. Devices used by the positioning system are also in or attached to the streamers.

Figure 4.21 shows, diagrammatically, the active part of a typical 24-bit ground or inwater system. Each "box" contains N channels of recording capacity, where N ranges from 1 to 8, depending on the manufacturer. Each channel has the same active components with a common output to the Line cable.

The first component is the preamplifier or "preamp". A magnifying glass (Fig. 4.22) makes things look bigger by a fixed amount, depending on its power. Similarly, the preamp increases input signal by a fixed amount that can be selected by the instrument operator before the recording starts. Typical choices available are: 0 dB (no increase), 12 dB ($\times$4), 24 dB ($\times$16), 36 dB ($\times$64), and 48 dB ($\times$256). Preamplifiers must be designed to introduce very low levels of instrument noise to maintain good signal-to-noise ratios.

The filters are analog filters, composed of resistors, capacitors and unity-gain amplifiers. There are two basic types of such filters: low-cut (high-pass) and high-cut (low-pass). Low-cut filters attenuate frequency components **below** a certain *low*

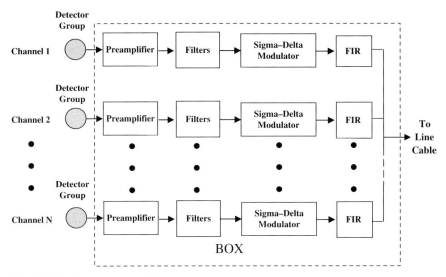

Fig. 4.21 Typical 24-Bit recording system

Fig. 4.22 Preamplifier function

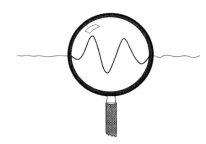

CHARACTERISTICS -

- FIXED MAGNIFICATION (GAIN)
- LOW NOISE GENERATION

cutoff frequency. High-cut filters attenuate frequency components **above** a certain *high cutoff frequency*. Modern seismic instruments have low-cut filters that may be used to attenuate low frequency noise. The responses of low cut and high cut filters are shown at the top of Fig. 4.23.

High-cut filters are not used in modern systems but another type, called a *notch* filter. A notch filter is a combination of low-cut and high cut filters with the low cutoff frequency higher than the high cutoff frequency. The resultant response is shown at the bottom of Fig. 4.23. Notch filters are mostly used to attenuate noise from electric power lines often referred to as highline noise. In the United States Of America, highline noise is at 60 Hz. In most other countries it is at 50 Hz.

Figure 4.24 is a functional diagram of a Sigma-Delta modulator. In this representation, the electronic switch opens and closes very rapidly. In some systems switching speed is 256,000 times per second. Each time the switch closes it presents a sample of the analog signal to the comparator. If the sample voltage is positive,

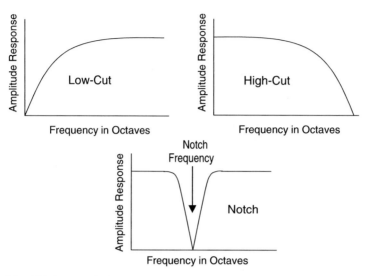

Fig. 4.23 Low cut, high cut, and notch filters

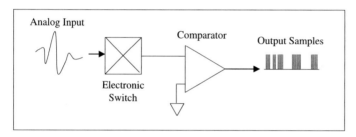

Fig. 4.24 Sigma delta modulator functional representation

the comparator outputs a pulse that represents the digit 1. If the sample voltage is negative, the comparator outputs a pulse that represents the digit 0.

A Sigma-Delta modulator is a 1-bit analog to digital converter. (A bit stands for binary digit and only 0 and 1 are used to represent binary numbers.) The analog (continuous) signal from the filters is input to the Sigma-Delta modulator and a string of pulses that represent the digits 0 and 1 are output. That is, the analog signal is sampled with the samples represented by binary (base 2) numbers. The sampling is done at a very rapid rate. This avoids a form of distortion called aliasing.

The Finite Impulse Response unit does three stages filtering and resampling. See Fig. 4.25. Stage 1 applies a weighted average or sinc function (sin x/x) to the output of the Sigma-Delta modulator, a string of 1-bit samples. It outputs data values of greater precision (12 bits). Resampling to an intermediate sample period is also done. For example, with 256,000 samples/sec as input, a 32:1 resampling rate delivers 8000 samples/sec as output. Stage 2 filtering applies a weighted average or sinc function (sin x/x) to the Stage 1 output that produces a very precise measurement of the sampled voltage (24-bit data word). Resampling that output data at twice the

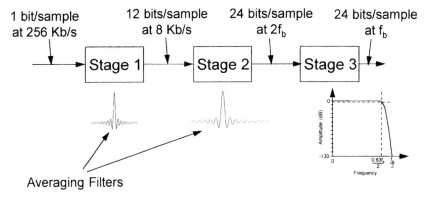

Fig. 4.25 Finite impulse response (FIR) filter

desired rate is also performed in Stage 2. Stage 3 does final bandwidth shaping by applying a high cut filter that is −3 dB at 83% of the Nyquist and −130 dB at the Nyquist frequency (half the final sample rate). Stage 3 then resamples by 2:1 to output data at the desired rate. Figure 4.26 illustrates resampling done in the FIR.

Major components of a truck-mounted or onboard system include:

- System Control Unit
- Line/Streamer Interface Modules
- Correlator/Stacker Module (truck-mounted, only)
- Operator Console Modules
- Tape Transports
- Printers and/or Cameras

The system control unit controls the flow of data from the line interface unit/streamer interface modules and to the tape transport units, as well as the functions of

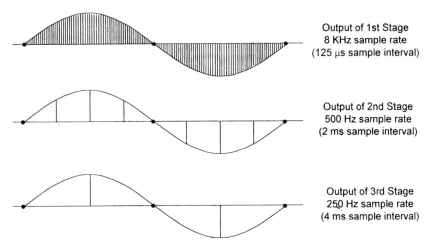

Fig. 4.26 Resampling in the FIR

peripheral devices such as cameras. It also interfaces with the navigation system and the airgun controller in marine operations to assure that recording begins and the airgun array fires at the right location. Streamer interface modules function much the same way as the ground system line interface modules. However, streamer interface modules are located onboard the vessel rather than in the streamers.

The correlator/stacker module is used with Vibrators in land operations. It performs crosscorrelation between the filtered pilot sweep and the raw vibrator records and vertical stacking (summing of like traces in a set of records taken at the same shotpoint).

The system operators use the operator console modules to input recording parameters and to monitor recording system performance. Data received from the system control unit go to the tape transport control units for formatting and then to the tape transports for recording on magnetic tape. Printers and/or cameras are used to produce "monitor records" (hard copy or visual versions of data recorded on tape), status reports, etc.

Some definitions of terms are required before continuing the discussion of recording. (See Table 4.3). Dynamic Range is the ratio of the largest signal that can be recorded to the root-mean-square (rms) value of instrument noise. Instantaneous Dynamic Range is the ratio of the largest signal that can be satisfactorily recorded to the smallest signal that is detectable, at any one time. System Dynamic Range is the ratio of the largest signal that can be recorded to minimum rms value of instrument noise. Harmonic Distortion is the ratio of the sum of the recorded amplitudes of second and higher order harmonics to amplitude of the input fundamental. (Harmonics are integral multiples of the input or fundamental frequency.) Crossfeed is the ratio of the amplitudes of electromagnetically induced signal in one channel to the amplitude input to another channel.

The most important of these specifications is instantaneous dynamic range. This is because the recorded data includes noise and signal superimposed on each other. Seismic processing is able to enhance data if the seismic signal is recorded within the instantaneous dynamic range of the recording system. Seismic signals may have an amplitude range of 120 dB or more when considering the amplitudes of direct arrivals and ground roll versus those of deep, faint reflections.

Seismic data are recorded on magnetic tape, which is a strip of plastic coated with iron oxide particles. Data are retained by magnetization of iron oxide particles by *write heads*. Each writes head magnetizes a particular portion of the tape as it passes under it. This is called a *tape track*.

Table 4.3 24-Bit system specifications

Specification	Value (dB)
Dynamic Range	120
Harmonic Distortion	106
Crossfeed	110+
Instantaneous Dynamic Range	120
Total Dynamic Range	140

   Recording techniques include both analog and digital. In analog recording data are represented by varying magnetization intensity along each track; one track per channel. In digital recording small, discrete areas of the tape are magnetized in one direction or the opposite direction. There is no connection between channel and track in digital recording. Digital recording of seismic data was introduced in the late 1960s and rapidly replaced analog recording. Analog recording systems have much smaller dynamic range than digital and analog data cannot be directly input to digital computers for processing.

   The two directions of tape magnetization in digital recording must be used to represent numbers so raw data can be processed and made interpretable. Therefore, changes in tape magnetization are used to represent the digits 0 and 1. The number system that uses only the digits 0 and 1 is called **binary**.

   To understand binary numbers, first consider the more familiar decimal system. Both binary and decimal systems use something called positional representation. For example, the decimal number 5763.41 is $5000 + 700 + 60 + 3 + 0.4 + 0.01$ or $5 \times 1000 + 7 \times 100 + 6 \times 10 + 3 \times 1 + 4 \times 0.1 + 1 \times 0.01$. This can also be written as $5 \times 10^3 + 7 \times 10^2 + 6 \times 10^1 + 3 \times 10^0 + 4 \times 10^{-1} + 1 \times 10^{-2}$. In other words, the position of a digit with respect to the decimal point indicates the power of 10 by which it is multiplied.

   Similarly, in binary numbers the positions of the 0s and 1s indicate the powers of 2 by which they are multiplied to represent a number. For example, the binary number 110111001.01011 represents $1 \times 2^8 + 1 \times 2^7 + 0 \times 2^6 + 1 \times 2^5 + 1 \times 2^4 + 1 \times 2^3 + 0 \times 2^2 + 0 \times 2^1 + 1 \times 2^0 + 0 \times 2^{-1} + 1 \times 2^{-2} + 0 \times 2^{-3} + 1 \times 2^{-4}$. Converting this to decimal gives $1 \times 256 + 1 \times 128 + 0 \times 64 + 1 \times 32 + 1 \times 16 + 1 \times 8 + 0 \times 4 + 0 \times 2 + 1 \times 1 + 0 \times 0.5 + 1 \times 0.25 + 0 \times 0.125 + 1 \times 0.0625 + 1 \times 0.03125$ or decimal 441.34375.

   Table 4.4 compares counting from 1 to 10 in decimal and binary. Note how many more digits are required to represent the same numbers in binary compared to decimal. If it were desired to print out exactly what was recorded in a seismic digital record (a tape "dump") a lot of 0s and 1s would have to be printed. A conversion to decimal would not allow the interpretation of each binary digit or "bit" value. Hexadecimal numbers allow this to be done by "compressing sets" of four bits into only two hexadecimal digits.

Table 4.4 Comparison of decimal and binary numbers

Decimal	Binary
1	1
2	10
3	11
4	100
5	101
6	110
7	111
8	1000
9	1001
10	1010

Table 4.5 Decimal, binary, and hexadecimal numbers

Decimal	Binary	Hexadecimal
0	0000	0
1	0001	1
2	0010	2
3	0011	3
4	0100	4
5	0101	5
6	0110	6
7	0111	7
8	1000	8
9	1001	9
10	1010	A
11	1011	B
12	1100	C
13	1101	D
14	1110	E
15	1111	F

As shown in Table 4.5, the hexadecimal system requires 16 digits, so the digits 0 through 9 are supplemented with the letters A (10), B (11), C (12), D (13), E (14), and F (15) are used to represent the "extra" digits required by the hexadecimal system. Note that the letters are always upper case. It can be seen from Table 4.5 that one hexadecimal digit represents four bits. Thus, in tape dumps a **byte** consisting of eight bits can be represented by two hexadecimal digits. For example if the eight bits of a particular byte are 11010100, it can be written in hexadecimal as D4. To see this first separate the byte into two groups of four bits, 1101 and 0100. In Table 4.5 the hexadecimal digit corresponding to 1101 is D and the one corresponding to 0100 is 4.

Another way of writing numbers in digital systems is binary-coded-decimal (BCD). The digits 0–9 are represented by four bits, as shown in Table 4.6.

Only 0s and 1s can be written on a digital tape, not + and –. Some other way to represent negative numbers is required. There are two methods of writing digital negative numbers – *one's complement* and *two's complement*. The two methods are demonstrated using the binary equivalent of decimal 873. The one's complement is formed by changing all 0s to 1s and vice versa. The two's complement is formed by adding a 1 to the least significant bit (the rightmost bit).

Table 4.6 Decimal to binary-coded-decimal (BCD)

Decimal	BCD	Decimal	BCD
0	0000	5	1001
1	0001	6	1010
2	0010	7	1011
3	0011	8	1100
4	0100	9	1101

$$+873_{10} = 000001101101001$$
$$-873_{10} = 111110010010110 = \text{One's complement}$$
$$\underline{\qquad\qquad\qquad +1 \qquad\qquad}$$
$$= 111110010011111 = \text{Two's complement.}$$

Computers can only perform addition! Multiplication is done by repeated addition. Division is done by repeated subtraction. To subtract, add the *complement* of the number to be subtracted.

Shown below is the subtraction of decimal 873 from itself, using the one's complement,

$$873_{10} = 000001101101001$$
$$-873_{10} = +111110010010110$$
$$\underline{= 0 \quad = 111111111111111}$$

The result of all 1s this does not, intuitively, seem to be a "right" answer but it works; the all 1s is sometimes called "negative zero".

Subtraction of decimal 873 from itself using the two's complement is shown below.

$$873_{10} = 000001101101001$$
$$-873_{10} = +111110010011111$$
$$\underline{= 0 \quad = 1|000000000000000}$$

The 16th bit created by the carry of one in adding the two numbers "overflows". That is, since, in this example, the register can only hold 15 bits, the 16th bit does not appear. Just 15 zeros are present. This result seems more reasonable. However, either complement yields valid results but which is being used must always be stated.

Figure 4.27 is a seismic digital tape schematic. Close to the start of a magnetic tape there is a physical, magnetic marker, called the *Beginning Of Tape* or *BOT*. This BOT must be sensed by the tape transport before recording of any data can start. Near the end of the tape is another physical, magnetic marker, called the *End Of Tape* or *EOT*. If the tape transport detects the EOT it evaluates remaining tape as to whether there is enough room on the tape to continue recording. If there is not enough tape remaining, the tape stops and the operator are signaled to change tapes.

A magnetic tape record is not the same as a seismic record. A *magnetic tape record* is a set of data terminated by an internal record gap (IRG). A *file* consists of one or more records separated from other files by an End-of-File (EOF) code (all 1s). A file is the largest division on tape.

A *byte*, which is the basic index of the tape format, consists of eight data bits. A *parity bit* is appended as a quality control method in magnetic tape recording. The parity bit is recorded as either a 0 or a 1 to assure that the number of logical 1s written on tape is odd (odd parity) or even (even parity). Seismic exploration

Fig. 4.27 Tape schematic

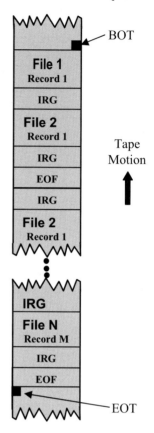

mostly uses odd parity, which is illustrated below. (Note: S = sign bit, MSB = most significant bit, LSB = least significant bit, and PAR = parity bit,)

Data Value: 0101010 (Three 1s)

Parity Bit Value: 0

$$\begin{matrix} M & & L & P \\ S & & S & A \\ S & B & & B & R \end{matrix}$$

Value Written to Tape: 0 0 1 0 1 0 1 0 0 (Three 1 s)

Data Value: 0110 0101 (Four 1s)

Parity Bit Value: 1

$$\begin{matrix} M & & L & P \\ S & & S & A \\ S & B & & B & R \end{matrix}$$

Value Written to Tape: 0 1 1 0 0 10 1 1 (Five 1 s)

Word is a computer term that is important when doing "tape dumps". Dumps are indexed by words. Word size varies from computer to computer.

Multiplexed data are written on tape in the order they are sampled. For simplicity, consider the recording of 20 ms of data at 4ms sample intervals in an 8-channel system. This will result in six samples for each channel. Let the initial samples recorded in channels 1 through 8 be A_{01}, A_{02}, A_{03}, A_{04}, A_{05}, A_{06}, A_{07}, and A_{08}. Similarly, let the second samples recorded in channels 1 through 8 be A_{11}, A_{12}, A_{13}, A_{14}, A_{15}, A_{16}, A_{17}, and A_{18}. Following the same pattern, the sixth samples recorded in channels 1 through 8 are A_{51}, A_{52}, A_{53}, A_{54}, A_{55}, A_{56}, A_{57}, and A_{58}. In multiplexed form the samples are read from the tape in the order:

$$A_{01}, A_{02}, A_{03}, A_{04}, A_{05}, A_{06}, A_{07}, A_{08},$$
$$A_{11}, A_{12}, A_{13}, A_{14}, A_{15}, A_{16}, A_{17}, A_{18}.$$
$$A_{21}, A_{22}, A_{23}, A_{24}, A_{25}, A_{26}, A_{27}, A_{28},$$
$$A_{31}, A_{32}, A_{33}, A_{34}, A_{35}, A_{36}, A_{37}, A_{38},$$
$$A_{41}, A_{42}, A_{43}, A_{44}, A_{45}, A_{46}, A_{47}, A_{48},$$
$$A_{51}, A_{52}, A_{53}, A_{54}, A_{55}, A_{56}, A_{57}, A_{58}.$$

It would be extremely difficult to process data in the multiplexed form. Processing requires the application of mathematical operations to all samples of a channel. Thus. multiplexed data have to be *demultiplexed*, that is data are re-ordered to be written in channel order, and i.e., all samples for the same channel are written together ordered by sample times. Using the above example, for demultiplexed data samples are read from the tape in the following order:

$$A_{01}, A_{11}, A_{21}, A_{31}, A_{41}, A_{51},$$
$$A_{02}, A_{12}, A_{22}, A_{32}, A_{42}, A_{52},$$
$$A_{03}, A_{13}, A_{23}, A_{33}, A_{43}, A_{53},$$
$$A_{04}, A_{14}, A_{24}, A_{34}, A_{44}, A_{54},$$
$$A_{05}, A_{15}, A_{25}, A_{35}, A_{45}, A_{55},$$
$$A_{06}, A_{16}, A_{26}, A_{36}, A_{46}, A_{56},$$
$$A_{07}, A_{17}, A_{27}, A_{37}, A_{47}, A_{57},$$
$$A_{08}, A_{18}, A_{28}, A_{38}, A_{48}, A_{58}.$$

When digital recording (early-to mid-1960s) began, many different tape formats were developed. To, facilitate tape exchange and bring order to the industry, the Society of Exploration Geophysicists (SEG) adopted standard formats. A brief summary follows:

1967 – *SEG A* and *SEG B* (field data, multiplexed), and *SEG X* (data exchange, demultiplexed)

1972 – *SEG C* (field data, multiplexed) introduced to accommodate IFP ecorders

1975 – *SEG Y* (demultiplexed) introduced as new data exchange format to accommodate computer field equipment and newer processing hardware.

1980 – *SEG D* (multi-purpose, multiplexed or demultiplexed, details in the header) introduced to accommodate further advances in data acquisition and processing. SEG D was revised in 1994 to accommodate other developments, including 24-bit recording.

Acquisition Methodology

Variations in seismic data acquisition methodology depend upon whether 2-D or 3-D data are to be acquired and whether the environment in which data are collected is land, marine, or ocean-bottom. Since 2-D methods were introduced first, they will be discussed first

2-D Acquisition

Figure 4.28 shows desired 2-D line geometry. Sets of receiver groups (the *spread*) are laid out along the lines and sources shot into them. After one is shot into, the spread and source are moved along the line to provide the desired subsurface coverage. In the land environment natural and cultural obstacles usually prohibit long straight lines such as shown in Fig. 4.28. Only in a few areas, such as deserts, can long straight lines be shot on land. A greater variety of geometrical relationships between sources and receivers are, however, possible in land operations than in other environments.

The geometrical relationship usually desired is called the *off end spread*, shown in Fig. 4.29. Here all receiver groups are on one side of the source. If the source is at the end of the receiver groups that is in the direction of progression along the

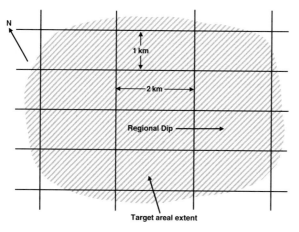

Fig. 4.28 Typical 2-D layout

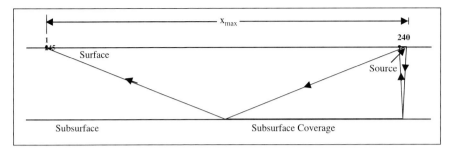

Fig. 4.29 Off end spread

line, the source is said to be "pulling the spread". If the source is opposite to the direction in which the shooting the line progresses, the source is said to be "pushing the spread".

Figure 4.30 shows a *split spread*. Since there are an equal number of receivers on each side of the spread it is a *symmetric split spread*. The split spread shown in Fig. 4.31 is an *asymmetric split spread* because there are more receivers on one side of the source than the other.

As indicated above, off end spreads are preferred. This is because all energy travels from source to receiver in the same direction. However, the number of groups

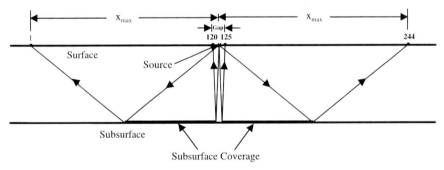

Fig. 4.30 Symmetric split spread

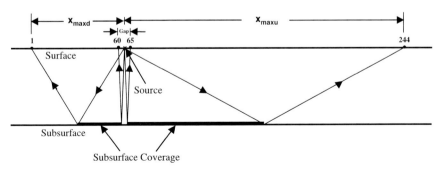

Fig. 4.31 Asymmetric split spreads

and the group interval required to yield desired results may make the maximum offset too large. Split spreads can be used to resolve this problem.

Symmetric split spreads are probably used more often than asymmetric split spreads. Asymmetric split spreads do, however, provide a couple of advantages. While the maximum offset should NEVER exceed the depth to the primary target, optimum velocities are obtained when the maximum offset is equal to or just less than depth to the primary target. An asymmetric split spread allows this situation in one direction and this is really a good thing.

If at all possible, it is desirable to have the receivers up-dip from the source. With off end spreads on land this is easily accomplished. With split spreads the source is always up-dip from some receivers. Figure 4.32 shows why this is not a good thing.

Ray paths from the source to the receiver for the far offset group in the up-dip direction are shown on the top left of Fig. 4.32. Note that the wave front incident on the receiver array is not quite horizontal. This results in a slight delay in the signal reaching successive receivers across the receiver array, as indicated at the bottom left. The bottom left of Fig. 4.32 represents the signals from individual detectors and the array output. (All receivers in the array are electrically connected so the array output is the sum of the individual receivers in the array). The delays result in a slight amount of attenuation and filtering.

In the down-dip direction (right side of Fig. 4.32) the wave front incident on the receiver array departs considerably from horizontal and delays in signal arrival at successive receivers are much larger than in the up-dip case. As shown at the lower right, there is much more attenuation and signal distortion.

Marine operations do usually allow long straight lines, although obstacles such as drilling platforms may impose some restrictions and/or special techniques. In marine 2-D it is assumed that the streamer is straight and follows the boat path.

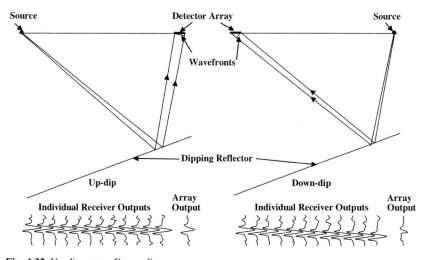

Fig. 4.32 Up-dip versus Down-dip

Fig. 4.33 Streamer feathering

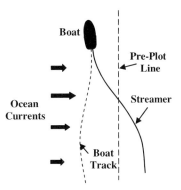

This is rarely, if ever, the case. As shown in Fig. 4.33, differences in ocean currents and boat turns cause the streamer to feather and bow into, usually, arced shapes. The boat must steer a path that places receivers as close as possible to the desired position (the pre-plot line). This means that receivers fall along an area rather than along a line resulting in some smearing of data. This is a problem for marine 2-D but in 3-D it can be made an advantage if the reflection points that are scattered over an area can be accurately located.

In single-boat marine acquisition, spreads must be off end with the source pulling the spread. Efficient acquisition techniques require that some lines be acquired with shooting in the up-dip direction and some in the down-dip direction. Doing otherwise would require long boat and streamer turnarounds that entail much time in which no data acquisition is done.

Figure 4.34 illustrates the ocean bottom cable (OBC) technique. Cables are laid out on the ocean floor. Receiver groups include both hydrophones and geophones. Often, three-component geophones are used. Separate recording and shooting boats are required. Airgun arrays are used, as in marine operations. The use of geophones and the need to correct for differences in receiver group depth variations are similar to land operations. Unique features of OBC operations are that receiver group locations must be determined from direct and refracted arrivals at the groups and scalars are needed to combine geophone and hydrophone data.

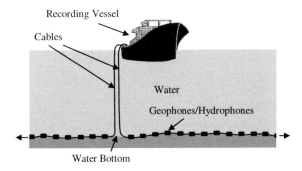

Fig. 4.34 Ocean bottom cable (OBC) system

The following acquisition parameter values must be determined before an acquisition program can start:

- Line parameters
    - Number and orientation of lines
    - Line spacing
    - Line lengths
- Source parameters
    - For explosives
        - Size (e.g., pounds of dynamite)
        - Number of holes
        - Hole depth
        - Pattern
    - For vibrators
        - Number and layout of source positions per source point
        - Number of units
        - Sweep type
        - Number of sweeps
        - Sweep length
        - Initial and final frequencies
    - For airguns
        - Number and sizes of guns
        - Array design
        - Number of arrays
        - Depth at which array is towed
- Spread parameters
    - Spread types
        - Off-end
            - Source pulling or pushing spread
        - Split-spread
            - Gap
            - Symmetric or asymmetric
    - Number of groups
    - Group Interval
    - Maximum and Minimum Offsets
- Fold

The last parameter, fold requires some explanation. Before the advent of digital recording of seismic data, a method called continuous subsurface coverage was used. Figure 4.35 illustrates this. Each spread provides 1/2 spread of subsurface coverage. Moving the spread 1/2 spread length between shots thus provides continuous coverage of the subsurface below the line.

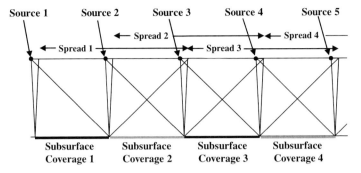

Fig. 4.35 Continuous subsurface coverage

When digital recording was implemented, it became practical to use a method called *multi-fold shooting*. Figure 4.36 shows one type of this, called four-fold shooting. Here the spread is moved 1/8 of a spread length between the shots. Subsurface coverage from the second shot overlaps 75% with the first. Subsurface coverage from the third shot overlaps 75% with the second and 50% with the first. Subsurface coverage from the fourth shot overlaps 75% with the third, 50% with the second and 25% with the first. The pattern continues with each subsequent shot to the end of the line.

As a result of the overlapping of subsurface coverage, the first 1/4-spread length of subsurface coverage is recorded only from the first shot. This is called *single-fold* or *1-fold*. The second 1/4-spread length of subsurface coverage is recorded from both the first and second shots. This is called *2-fold*. The third 1/4-spread length of subsurface coverage is recorded from the first, second, and third shots. This is called *3-fold*. The fourth 1/4-spread length of subsurface coverage is recorded from the first, second, third, and fourth shots. This is called *4-fold*. Subsequent shots continue

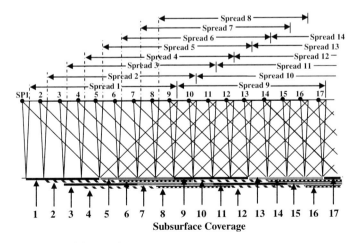

Fig. 4.36 Four-fold shooting

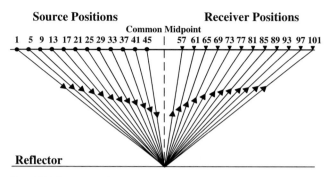

Fig. 4.37 Common midpoint ray paths

the pattern of having 1/4 –spread lengths of subsurface coverage in common, so 4-fold coverage continues through the fourth from the last 1/4-spread coverage on the line. Fold then reduces to 3-, 2- and 1-fold over the last three 1/4-spread lengths of subsurface coverage. Folds as high as 24, 30, or even 60 are used today.

To see the utility of multifold shooting, consider the recording of a single reflection on multiple records. Figure 4.37 illustrates this for 12-fold shooting. Twelve different traces from 12 shot records record reflections from the same or almost the same position vertically below the midpoint between the sources and receivers. The source-to-receiver offsets are different for each trace so all will have different amounts of normal moveout (NMO). This provides a means of determining the velocity to use in NMO correction. After correction for NMO the traces can be combined in a process called common midpoint (CMP) stacking that enhances signal-to-noise ratio and attenuates multiple reflections.

In the discussion above, the expression "move the spread" may give the impression that, in land operations, all receiver groups and cables are picked up and moved a certain number of group intervals along the line. This is not the case. The usual practice is to lay out along the line many more receiver groups on the ground than are required in one spread. All are connected by cable to the instrument truck. The instrument operator can select the particular receiver groups appropriate to each shot. Receiver groups and cables that were used in earlier shots can be picked up and moved ahead. The instrument truck can stay in one location for a great many shots.

3-D Acquisition

Nearly all seismic surveys are now 3-D. Very little 2-D shooting is done today because of the shortcomings of 2-D such as:

- Distortion of the image of geologic structure
- Inadequate subsurface sampling to define small-scale geologic features

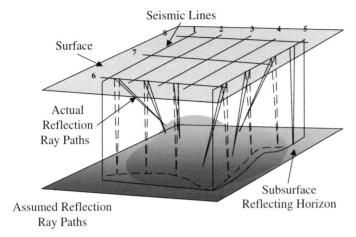

Fig. 4.38 Actual and assumed ray paths from a subsurface horizon – 2-D case

Figure 4.38 illustrates the distortion problem that results because the basic premise of 2-D shooting is that all reflections originate vertically below the seismic line. Here the reflecting horizon has a dome structure. A few reflection ray paths are shown for 2-D lines 5 and 6. The solid lines indicate where reflections actually occur and dashed lines indicate the assumed ray paths in the vertical plane passing through the lines. The solid curved line below line 5 is a true cross section of the structure. The dashed curve line is the apparent cross section. The true structure in the vertical plane passing through line 6 is flat. The assumption that the reflection ray paths were in this vertical plane produces the erroneous cross section shown by a curved dashed line below line 6. In general, interpretation of 2-D data shows a reflecting surface that differs from the actual in position and has less dip than the actual reflecting surface.

For reflectors that are flat or dip very little the 2-D assumption is not so bad. However such structures rarely provide adequate trapping capability to produce petroleum deposits that are commercially viable.

Figure 4.39 (a) shows a set of 2-D lines drawn over true depth contours that indicate subsurface geological structure. Figure 4.39(b) shows one of the many possible depth contour maps that could be drawn through the observed points of equal depth. This interpretation does not show the true structure. It is almost impossible to connect equal depth points correctly and some closed contours are completely missed by the 2-D grid. 3-D acquisition eliminates both these problems.

Figure 4.40 is an example of the preferred 3-D geometry. Lines are much more closely spaced than in 2-D. Receiver lines are usually laid out only in the direction of maximum target dip. Source lines are usually perpendicular to receiver lines and spaced farther apart. Reflection points on the subsurface are spaced at half the group interval in the direction of the receiver lines (the *inline* direction) and half the line interval in the direction of the source lines (the *cross-line* direction). This provides much greater spatial sampling and far less interpretational ambiguity.

Fig. 4.39 Poor subsurface
sampling from 2-D Data.
(**a**) True depth structure.
(**b**) Interpretation based on
points of equal depth

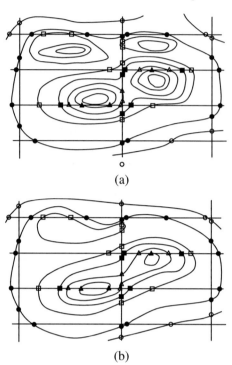

(a)

(b)

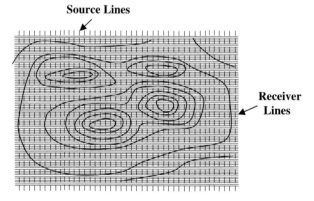

Fig. 4.40 3-D layout example

As in 2-D, it is rarely possible to implement long straight lines in land operations
but every attempt is made to obtain the desired subsurface sampling over the target.
In marine 3-D operations, recording boats pull multiple streamers – as many as
12. This faciltitates obtaining the dense subsurface coverage needed but introduces
the need to accurately determine the position of every hydrophone group in every
streamer for every shot. In OBC operations multiple cables are laid out but their
spacing cannot be as precisely controlled as in land operations.

In 2-D land operations the combination of receiver groups connected to the instruments to record a single record is called a spread. Only one source is shot into a spread, after which the spread is moved forward along the line. Reflection points fall along lines.

Figure 4.41 illustrates a 3-D land operations procedure particularly well adapted to the use of vibrators. Eight receiver lines are laid out but only six are active at a time. The total length of the six lines is called a *swath*. The portion of the swath enclosed by a rectangle at the lower right of the Figure is called a *patch*. These are the receiver groups used for the first shot. The source position for this patch is circled. The subsurface reflection points are distributed over an area instead of along a line. The patch and source are moved up the swath in the direction shown. In some cases the source line is continued only to the point that allows recording from a full-sized patch. It is usually more efficient to extend the source line beyond this and to reduce the size of the patch accordingly. This allows full fold over a greater area. See Fig. 4.42.

When the first swath is completed, one or more receiver lines are moved laterally (*rolled*) such that there is overlap in surface and/or subsurface coverage. Figure 4.43 illustrates a one-line roll. This continues until all sources have been shot and the entire survey area covered.

There are many different approaches to acquisition of land 3-D data. Each method attempts to optimize efficiency and minimize costs. In most cases, occupying the same source position twice is to be avoided. The main concern is to obtain the desired subsurface sampling.

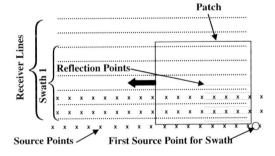

Fig. 4.41 Patch, swath, source point, and reflection points for the first patch

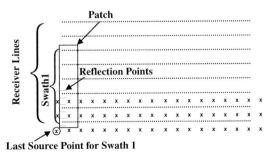

Fig. 4.42 Situation at the end of a swath

Fig. 4.43 One-line roll to
patch 2

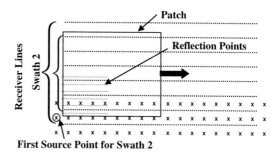

In Marine 3-D multiple streamers are towed behind the vessel. Since streamers rarely, if ever, are straight and follow the boat track, midpoints tend to be quite scattered with few, if any, truly common midpoints, some method must be developed to combine traces as needed for CMP Stack, velocity determination, etc. The approach taken is to divide the survey area into *bins* that are usually rectangular, as shown in Fig. 4.44. All traces whose midpoints fall in the same bin are considered common midpoint traces. (Note: bins are also used in land and OBC surveys.)

Marine 3-D requires the capability to accurately track the positions of the hydrophone groups in all streamers throughout the survey. Without precise knowledge of the streamers relative to the vessel, midpoint positions cannot be determined and assigned to the correct bins. A method of online binning (determining in which bins midpoints fall) is used to aid vessel steering to better meet objectives of subsurface sampling as well as onboard quality control processing.

The trend to smaller group intervals and shot intervals has required that two airgun arrays be used, to be fired alternately. This is because there is insufficient time for an airgun array to be fully charged between consecutive source points.

The increase in the number and length of streamers has improved subsurface sampling capability but has also caused line changes to be somewhat more difficult. When a line is completed not only the vessel but also all streamers must have their

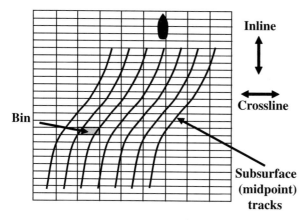

Fig. 4.44 Marine 3-D situation

directions changed. The turn must be wide enough and long enough to remove the effects of the turn from the streamers' positions.

The same parameters described for 2-D acquisition apply to 3-D. However, to obtain the best possible subsurface image requires consideration of the three areas shown in Fig. 4.45:

. Primary target
. Migration aperture
. Fold taper

It is desirable to have *maximum fold* over the entire area that encloses the target. In 3-D fold is the product of *inline fold* and *crossline fold*. Inline fold is the same as the fold discussed for 2-D. It arises from the overlap in subsurface coverage as the patch is moved along the swath. Cross-line fold results from rolling lines from one swath to the other. For example, in Fig. 5.43 there is a six-line swath and a line roll of one. This gives a cross-line fold of 3, since:

$$XFOLD = \frac{NL}{2 \times LR} \qquad (4.1)$$

where XFOLD = crossline fold

NL = number in lines of the swath
LR = lines rolled

Inline fold can be calculated from:

$$IFOLD = \frac{NG}{2 \times GR} \qquad (4.2)$$

where IFOLD = inline fold

NG = number of groups in one line of a patch
LR = groups moved per line from one patch to the next along the swath

If there are 256 groups in one line of a patch and the patch is moved eight groups, inline fold is $256/(2 \times 8) = 16$. If the crossline fold is 3, the maximum total fold is $3 \times 16 = 48$.

As noted in the discussion of 2-D fold, maximum fold is not obtained immediately but must be built up as shooting progresses along a line. Similarly, at the end of the line, fold decreases from the maximum to single fold. This is called fold taper. Crossline fold also builds up to maximum and decreases to single fold at the other side of the survey. Thus, it is necessary to extend the lines at the beginning and end of the swaths and to add extra lines on the edges to obtain the desired fold.

Migration is a process that moves data from apparent subsurface positions to true positions. It requires a great deal of data from outside the target zone. This is due to the fact that an apparent position frequently lies outside the target area but its true position is within the target area. (See Fig. 4.45). Thus, the area around the target must be expanded to include additional lines and additional receiver groups to ensure that all relevant data are present for the migration process. This is called the *migration aperture*.

Fig. 4.45 Areas to consider
in determining number and
lengths of 3-D lines

All data within both the target area and within the migration aperture should be at maximum fold. This means that the fold tapers must be added to the migration aperture, not just to the target area.

Vertical Seismic Profiling (VSP)

The VSP concept is rather simple. A geophone is lowered to the bottom of a borehole. A seismic signal is generated at, or near, the surface. The signal received by the borehole geophone is recorded. The borehole geophone is raised by a predetermined amount and the process is repeated until the shallowest depth of interest is reached. The result is a VSP record comprised of "traces" recorded at various depths in the well. Figure 4.46 shows the basic VSP setup and concept. (Note: The distance between the source and the borehole is small enough for raypaths to be nearly vertical, smaller than it appears in Fig. 4.46.) In surface seismic profiling the source and the receivers are on the surface, aligned more or less horizontally. In VSP the geophone is aligned more or less vertically.

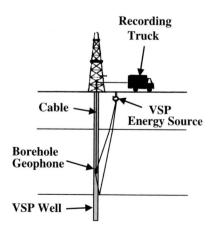

Fig. 4.46 VSP concept

As shown in the figure, the borehole geophone responds to both upgoing and downgoing seismic events. In conventional seismic surveys only upgoing seismic signals reflected from subsurface reflectors can be recorded by geophones at the surface. There is a superficial resemblance between VSP and a velocity survey because the source and receiver geometry is the same for both techniques. However, VSP and velocity surveys differ in two ways:

1. Depth increments between geophone recording depths in VSP are 15–40 m, while in the velocity surveys depth intervals may be a few hundreds of meters.
2. Only first break times are of real interest in a velocity survey, while the upgoing and downgoing events as well as first breaks are significant in a VSP survey.

Field Equipment and Physical Environment Requirements

Obviously, a VSP survey requires a borehole. An energy source is needed to generate a seismic signal. A downhole geophone is needed to detect the signal and convert it to an electrical pulse, and a recording system is required to record the events reaching the downhole geophone. Other equipment and physical factors that may be involved in VSP data acquisition will be discussed later. Design of the equipment is beyond the scope of this book and will not be discussed. There are suitable references in the bibliography at the end of this chapter for those interested in learning more about the equipment design.

An acceptable borehole must exist to run a VSP. Factors that should be considered in selecting a hole include:

- Hole deviation, i.e. – departures from vertical. VSP surveys are more economical and present fewer interpretation problems if they are conducted in vertical holes. A deviated hole creates uncertainty about the position of the downhole geophone relative to the energy source. If the source is moved to several different locations during the data acquisition phase, determining the position of the downhole geophone relative to the energy source is further complicated. An accurate deviation survey should be run in deviated holes to alleviate this problem. Interpretation problems are more common and more severe on VSP's in offshore wells because they are usually highly deviated. A VSP in a deviated well does, however, allow the subsurface beneath the borehole to be imaged laterally with great resolution. If the intent is to identify depths and one-way times for primary reflections, a vertical hole is the better choice. A VSP can be recorded more quickly and easily and with more accurate results in a vertical well.
- Casing and cementing. A cased hole is preferable for VSP recording because the geophone is protected from hole-caving and differential pressure problems. An uncased hole may have to be re-entered periodically for conditioning but a cased hole does not. Thus, survey time is not limited in a cased hole, Cemented casing is most desirable because there must be a medium between the casing and the borehole that is a good transmitter of seismic energy. Cement is the best medium.

- Borehole diameter. If the proposed hole is uncased, its diameter will likely vary and the borehole wall will be rough. This will affect clamping of the geophone to the formation, especially in large washouts where clamping may be impossible because the geophone-locking arm is too short to reach the borehole wall. To prevent such occurrences when an uncased well must be used, hole diameter should be measured with a caliper log. This allows selection of recording depths that avoid problems. Blair (1982) concluded that the seismic detector can be installed at any point on the circumference of a cylindrical borehole and still record the same particle motion, as long as the wavelengths of interest in the wavelet propagating past the borehole are greater than 10 times the circumference of the hole.
- Borehole obstructions. A cased hole may have obstructions that prevent the tool from reaching depths below them. The inability to reach these depths maybe critical to the survey. Checking for obstructions must be done before starting a VSP survey. Running a cheap tool, of the same diameter as the geophone, into the well to see whether obstructions exist before the VSP survey starts can do this.
- Borehole information. Complete interpretation of VSP data requires a suite of independent data that describes the physical properties of the formations around the borehole. A suite of logs, such as caliper, sonic, density, resistivity, and radioactive plus cores and drill cuttings taken from the well would satisfy this requirement. A cement bond log and measurements of the depths of all casing strings are required to be sure of the nature of the acoustic coupling between the VSP geophone and the formation.

The order of preference among the four common borehole environments for VSP data recording, is:

1. Cemented single casing
2. Uncased
3. Uncemented single casing, drilled sufficiently long ago that mud and cuttings in annulus are solidified
4. Recently cased and uncemented

Since VSP data is used to improve the interpretation of surface-recorded seismic data, the VSP should have the same wavelet and high-frequency content as the surface-recorded seismic data. This provides better correlation and tie between the two. If this is not possible, a match of the two wavelets must be done in the data processing stage. In conventional VSP recording the source location is relatively near the borehole so that signal raypaths will be nearly vertical.

VSP energy sources should satisfy the following:

1. Generate a consistent and repeatable shot wavelet. This is necessary to correlate equivalent characters of upgoing and downgoing wavelets throughout the vertical section over which data is gathered. Figure 4.47 illustrates this point.
2. Provide an output level that gives optimum response without overkill. It is *not* true that "the bigger the energy source, the better the response."

Fig. 4.47 VSP Energy source
wavelet

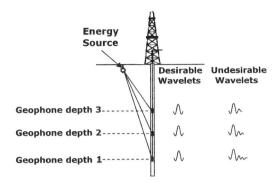

3. The increase in the number and amplitude of downgoing events must be greater
   than the gain of the amplitude of the upgoing events. This is because downgoing
   events are much stronger than the upgoing and as the output strength of the VSP
   source increases, more downgoing events will be created within numerous rever-
   berating layers in the near-surface part of the stratigraphic section. A decent VSP
   survey can often result when using an energy source of modest strength.

Surface energy sources used in VSP surveys are much the same as those used in
surface seismic surveys. They include dynamite, vibrators, air guns, and mechanical
impulse source.

Buried dynamite charges are widely used as the surface energy source for VSP
because of their effectiveness in producing seismic body waves. However, maintaining a consistent wavelet shape when shooting several tens of shots requires a
great deal of care. Vibrators are attractive for use in VSP work. A pilot sweep can
be designed and input to the ground that satisfies whatever resolution is required in
VSP recording. Sweep parameters such as number of units, length of sweep, and
number of sweeps can be selected that provides the desired signal-to-noise ratio.
Also, cross-correlation of Vibroseis sweeps enhances signal-to-noise ratio by discriminating against noise outside of the sweep frequency range. Coherent noise with
frequencies in the sweep bandwidth may present a problem but these can usually be
solved in the data processing stage. Wavelets produced by Vibroseis are repeatable
and consistent.

Air guns are, by far, the most widely used source in offshore vertical-seismic
profiling. An offshore vertical well allows many VSP objectives to be achieved.
The airgun can be placed at a fixed location near the wellhead suspended from a
work crane, making operations simple. The airgun can be operated from the high-
capacity air compressor that is standard equipment on the rig. Firing an airgun in this
arrangement may cause the rig to vibrate, but it will not cause any structural damage.
It is much safer than explosives for use on offshore rigs. Figure 4.48 illustrates the
arrangement described above.

Airguns can also be used as sources for onshore VSP. They are small and
portable, can be fired at intervals of a few seconds, and generate highly repeatable wavelets. The airgun must be submerged in water in order to function properly.
Figure 4.49 illustrates the airgun as an onshore VSP energy source.

Fig. 4.48 Airgun used as a
stationary energy source in
marine VSP surveys

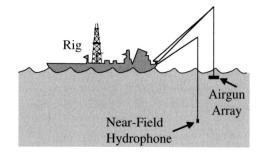

Fig. 4.49 Using the air gun as an onshore VSP energy source

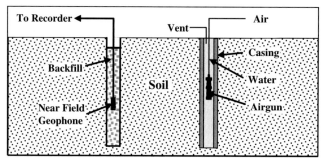

Mechanical impulse sources exist that can apply a vertical impulsive force to generate seismic energy. However, these sources should be tested for an area before being used.

There is a major difference in the shape, size, and construction of a geophone used for surface recording and a borehole geophone used to record a VSP survey, as shown in Fig. 4.50. A typical land geophone is about 10 cm long, has a diameter of about 3 cm, and weighs around 200 gm. By contrast, a downhole geophone is 3 m long, has a diameter of 10 cm, and weighs 100 kg.

The size of a downhole geophone results from its being within a massive housing that is designed to withstand the high pressures and temperatures found in deep wells. Also within this housing is the mechanical deployment system that anchors the geophone to the borehole wall and electronic amplifier and telemetry circuits.

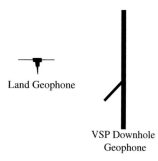

Fig. 4.50 Comparison
between surface and borehole
geophones

   The 24-bit recording systems used for surface seismic surveys will record down-hole geophone data and near-field monitor geophone responses with more than adequate resolution to capture high-resolution wavefronts. The near-field wavelet should be recorded in all marine VSP surveys. This is particularly true when using an energy source, such as an untuned airgun that creates a long wavelet. Signature deconvolution (see Chap. 5) can be used to compress the source wavelet and a recording of the near-field wavelet is needed for this purpose.

   Rayleigh waves, or ground roll, propagate along the earth's surface in all directions away from the energy source, interfering with the signal from deep reflectors in land exploration. These waves are undesirable signals and they prevent optimum imaging of stratigraphic, and structural conditions. However, the VSP does not record Rayleigh or Love waves, because they do not reach the depth of the geophone. Random noises are caused, in some wells, by formation irregularities, fluid movement behind the casing or in the well. There will be no further discussion of theses noises but there are other noises that deserve some discussion. These include:

- **Geophone coupling.** Poorly planted geophones cause some of the seismic noise recorded in surface geophones. Similarly, a poorly coupled downhole geophone produces undesirable noise. Figure 4.51 contrasts recordings from an unclamped (poor coupling) and clamped geophones.
- **Cable waves.** Propagation velocities of acoustic waves along cables that are used in VSP surveys depends on specific cable design but are in the range of 2,500–3,500 m/s.
  Because of this relatively high velocity, a cable wave can be the first arrival measured by the borehole geophone in shallow holes and low-velocity stratigraphic sections, If not correctly recognized, this wave can cause erroneous determinations of formation velocity. Cable-borne events are generally caused by wind or vibrating machinery. However, slacking the cable after the tool is locked downhole can reduce this type of noise as demonstrated in Fig. 4.52.
- **Resonance in multiple casing strings.** If one or more of the multiple casings is not being bonded to another string or to the formation it is difficult to obtain usable VSP data. In such cases, poor quality data are usually recorded near the surface where multiple casing is encountered. Processing can attenuate some of

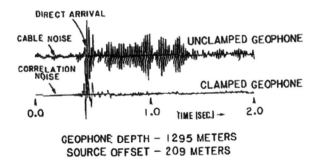

Fig. 4.51 VSP Geophone coupling

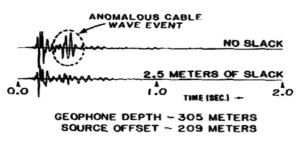

Fig. 4.52 Effect of cable slack on VSP signal

the noise patterns recorded, but a good cement bond between the strings is the key to good VSP data inside multiple casing strings.

- **Bonded and unbonded casing.** The shape of the wavelet recorded in cemented casings does not differ from one recorded in open hole. Since cement is the medium best capable of transmitting seismic energy from the formation to the geophone there is no the deterioration in data quality that is observed in un-bounded single casing.

- **Tube waves.** This is a disturbance that propagates along steel casings with a velocity of approximately 5.5 km/s, as reported by Gal'Perin(1974). See Fig. 4.53. Tube waves are among the most damaging noise patterns, because they are coherent noise. Summing multiple shots cannot reduce it. In fact, summing usually amplifies it, since its character is consistent for all records being summed. Careful data processing procedures may attenuate the tube wave effectively.

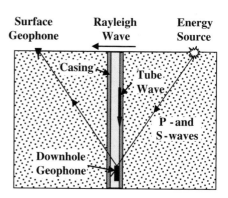

Fig. 4.53 Tube wave

VSP Field Procedures

Efficient field operations are essential because the longer a VSP survey is in progress, the more likely are equipment failures or borehole deterioration. Correct VSP field procedures should begin and end with a set of instrument tests to confirm that the recording system itself does not attenuate some signal frequencies, cross-feed has not been introduced into the data, and recording intervals are accurately determined.

A geophone tap test should be done. This determines the polarity of the geophone's output signal when its case moves in a specified direction in the borehole. This test is essential for correlation between VSP measurements and seismic data. It also serves as a check to make sure that the geophone is functioning before it is lowered into the borehole.

Energy input requirements for VSP are directly opposite those for velocity surveys. In the latter, energy input increase with depth may be appropriate to obtain good first breaks at all depths. In VSP downgoing and upgoing events, as well as first breaks, must be recorded. This may require more energy input when the geophone is at shallow depths so weak reflections from deep horizons can be adequately recorded. The energy required with the geophone at shallow depths may be two or three times as great as when the geophone is at the bottom of a deep hole.

Multiple shots and summing the resulting records is a preferable way to achieve the increase in energy. This will maintain wavelet consistency while increasing the size of the energy source may produce undesirable changes in wavelet character.

While the geophone is lowered into the well, test shots should be recorded at intervals of 300–500 m (1,000–1,700 ft). This allows selection of the appropriate recording parameters, e.g., source energy and number of shots to be summed to achieve adequate S/N ratio.

The VSP data is recorded as the borehole geophone is raised from the bottom of the hole to the surface. Shots should be at increments of a few meters. When the borehole geophone is around the target horizon, the depth increment is normally small, say100 ft (30 m) or less. The increment gets progressively longer as the geophone is pulled out of the hole. Multiple records are normally recorded at each geophone depth and vertically summed in the data processing stage.

Figure 4.54 shows a raw VSP record. The horizontal scale is in ft or m and the vertical scale indicates time in seconds. Each "trace" shows the events recorded when the geophone is at a particular depth in the hole.

The first strong amplitude events are called the *direct arrivals* or *downgoing wave*. Their travel times increase right to left, i.e. from shallow to deep. The downgoing waves are followed by downgoing multiples along the recorded VSP. Another event that can be seen is an *upgoing wave*. This event has a travel time trend of increase from left to right, i.e. – from deep to shallow. It is a mirror image of the downgoing events. These downgoing and upgoing waves must be separated from one another to allow useful information to be obtained from each type of waves.

Summary and Discussion

Seismic data acquisition requires careful planning based on technical and cost considerations. A specific primary target must be selected and documented so that acquisition parameters can be optimized. Equipment, personnel, and methodology must be selected to best meet plan objectives within the budget allotted. The potential returns from a successful survey must be sufficient to cover costs of the survey and development plus a reasonable profit.

Fig. 4.54 Vertical seismic
profile

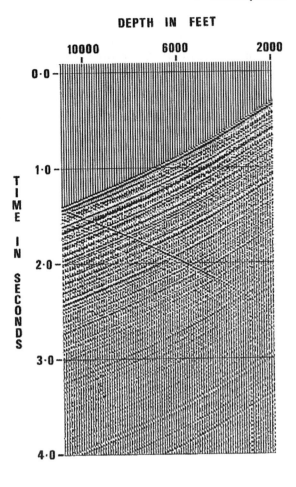

Before any seismic survey can begin it is necessary to obtain permission to operate in the area. This may require consent of individuals, corporations, or government agencies. All source and receiver positions must be determined with respect to a known point on the earth's surface. On land, not only positions but also elevations must be known. For OBC the equivalent of elevations, water depths, must be known.

Energy sources that best meet technical and cost requirements must be selected. On land, either vibrators or explosives are nearly always used. For marine and OBC surveys, airgun arrays are almost universally used.

On land, groups of geophones are connected to cables so that seismic energy produced by the source can be detected, converted to electrical pulses, and transmitted to the recording system. Modern recording instruments place parts of the recording system very near the geophones, within the cables to which the groups are connected. Other cables transmit the electrical impulses that have been converted to digital signals to elements of the recording system that monitor, format, and encode the data into magnetic form on magnetic tape cassettes.

In marine surveys groups of hydrophones are located within streamers that are towed behind the seismic vessel. Similar to land systems, parts of the recording system are located within the streamers near the hydrophone groups, as are conductors through which digital signals are transmitted to instruments onboard the vessel that monitor, format, and encode the data into magnetic form on magnetic tape cassettes.

3-D acquisition has practically supplanted 2-D. This is because it solves geophysical problems much better. Combined with advanced seismic processing packages, 3-D data provide high quality and high-resolution data that are used to locate new drilling prospects, delineate their boundaries, and provide information used to select drilling locations that optimize reservoir production. 3-D surveys have also been shown to be excellent tools in reservoir monitoring.

Vertical seismic profiling (VSP) has proved its value in applications to petroleum exploration and development. The cost of a VSP was substantial a few years ago but these days it is substantially less. The turnaround time is a few days and sometimes overnight in case of emergency. The survey is done routinely as any logging tool. In a vertical survey eight to ten levels per hour can be surveyed. In land surveys, perhaps six to eight levels per hour can be taken because it takes more time to inject the energy source. In offshore surveys, it may take more time, four to five levels per hour.

A VSP survey provides the geophysicist with seismic velocity, seismic time to geological depth conversion, and the next seismic marker. It provides the geologist with well prognosis. It will tell the engineer the location of the drilling bit or at what depth he can expect a high-pressure zone. If he can predict the high-pressure zone ahead of time, he can take action to head off problems. With minimal rig idle time, the survey is definitely more economical than a blowout.

The VSP plays an important role in borehole geophysics, reservoir characterization, and transmission tomography.

Exercises

1. How does depth of the charge affect recording of seismic data?
2. List two advantages and two disadvantages of using Vibrators as the energy source on land..
3. What is the "bubble effect" produced by single airguns? What is done to minimize it?
4. If the airgun arrays are at a depth of 6 m and the streamer is being towed at a depth of 10 m, at what frequencies would the ghost notches be seen? (Assume water velocity is 1500 m/s.)
5. Identify the natural frequency of the geophones from the amplitude responses shown below.

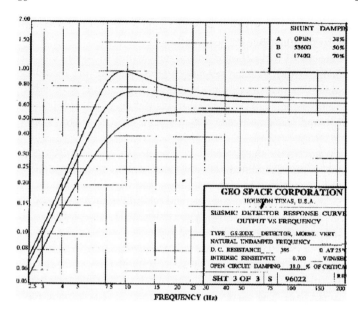

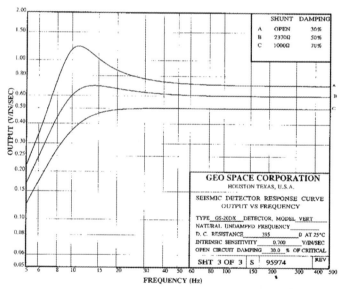

6. Shown below are dimensionless wavenumber responses for two linear arrays. How many elements are in each array?

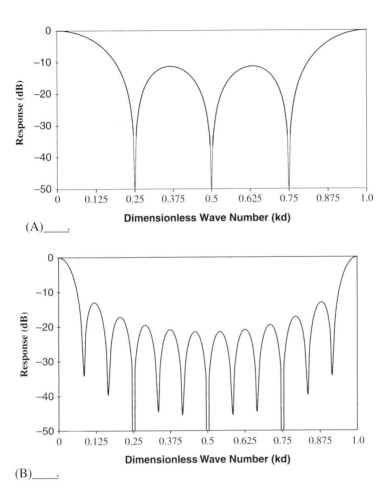

(A)_____.

(B)_____.

7. Show how decimal 583 is subtracted from decimal1947 in computers.
8. Sketch an off-end spread with the source pushing the spread.
9. A 192 group spread is to be shot at 24-fold. How many groups must be moved forward between shots to achieve this?
10. Theoretically, all lines should be shot from one direction. Why is this particularly impractical in marine seismic surveys?
11. List at least three significant differences between VSP and surface seismic acquisition.

Bibliography

Balch, A. H., M. W. Lee, J. J. Miller and R. T. Ryder. "The Use of Vertical Seismic Profiles in Seismic Investigations of the Earth." *Geophysics 47* (1982):906–918.

Balch, A. H. and M. W. Lee. "Some Considerations On the Use of Downhole Sources in Vertical Seismic Profiles." Paper presented at 35th Annual SEG Midwestern Exploration Meeting, (1982).

Blair, D. P. "Dynamic Modeling of In-Hole Mounts for Seismic Detectors." *Geo-phys. Journal of the Royal Astronomical Society 69* (1982):803–817.

Chapman, W. L., G. L., Brown, and D. W., Fair, "The Vibroseis System, a High Frequency Tool." *Geophysics 46* (1981):1657–1666.

Cheng, C. H. and M. N. Toksoz. "Elastic Wave Propagation in a Fluid-Filled Borehole and Synthetic Acoustic Logs." *Geophysics 46* (1981):1042–1053.

Cheng, C. H. and M. N. Toksoz. "Tube Wave Propagation and Attenuation in a Borehole." Paper presented at Massachusetts Institute of Technology Industrial Liaison Program Symposium, Houston, (1981).

Cheng, C. H. and M. N. Toksoz. "Generation, Propagation and Analysis of Tube Waves in a Borehole." Paper P, Trans. SPWLA 23rd Annual Logging Symposium, vol. I., and (1982).

Courtier, W. H. and H. L Mendenhall. "Experiences with Multiple Coverage Seismic Methods." *Geophysics 32* (1967):230–258.

Crawford, J. M., W. E. N. Doty and M. R. Lee. "Continuous Signal Seismograph." *Geophysics 25* (1960):95–105.

DiSiena, J. P. and J. E. Gaiser. "Marine Vertical Seismic Profiling" Paper OTC 4541, Offshore Technology Conference, Houston, TX, (1983), pp. 245–252.

Dix, C. H. "Seismic Velocities from Surface Measurements." *Geophysics 20* (1955):68–86.

Dobrin, M. B. *Introduction to Geophysical Prospecting.* New York: McGraw-Hill, (1960).

Douze, E. J. "Signal and Noise in Deep Wells." *Geophysics 29* (1964):721–732.

Gal¡ENT FONT=(normal text) VALUE=39¿'¡/ENT¿perin, E. I. "Vertical Seismic Profiling." *Society of Exploration Geophysicists Special Publications 12* (1974):270.

Gardner, D. H. "Measurement of Relative Ground Motion in Reflection Recording." *Geophysics 3* (1938):40–45.

Giles, B. F. "Pneumatic Acoustic Energy Source." *Geophysics Prospect 16* (1968):21–53.

Griffiths, D. H. and R F. King. *Applied Geophysics for Engineers and Geologists.* London: Pergamon, 1965.

Hardage, B. A. "An Examination of Tube Wave Noise in Vertical Seismic Profiling Data." *Geophysics 46* (1981):892–903.

Hardage, B. A. "A New Direction in Exploration Seismology is Down." *The Leading Edge 2,* no. 6 (1983):49–52.

Kearney, P. and M., Brooks, *An Introduction to Geophysical Exploration*, Blackwell Science Publication, Oxford, (1984):296pp.

Kennett, P., R. L. Ireson, and P. J. Conn. "Vertical Seismic Profiles—Their Applications in Exploration Geophysics." *Geophysics Prospecting 28* (1980):676–699.

Lang, D. G. "Downhole Seismic Technique Expands Borehole Data." *Oil and Gas Journal 77,* no. 28 (1979):139–142.

Lash, C. C. "Investigation of Multiple Reflections and Wave Conversions By Means of Vertical Wave Test (Vertical Seismic Profiling) in Southern Mississippi." *Geophysics 47* (1982):977–1000.

Lee, M. W. and A. H. Balch. "Theoretical Seismic Wave Radiation From a Fluid-Filled Borehole." *Geophysics 47* (1982):1308–1314.

Levin, F. K. and R. D. Lynn. "Deep Hole Geophone Studies." *Geophysics 23* (1958):639–664.

Marr, J. D. and E. F. Zagst. "Exploration Horizons from New Seismic Concepts of CDP and Digital Processing." *Geophysics 32* (1967):207–224.

Mayne, W. H. "Common Reflection Point Horizontal Stacking Techniques." *Geophysics 27* (1962):927–938.

Mayne, W. H. "Practical Considerations in the Use of Common Reflection Point Technique." *Geophysics 32* (1967):225–229.

Mayne, W. H. and R G. Quay. "Seismic Signatures of Large Air Guns." *Geophysics 36* (1971): 162–1173.

McCollum, B. and W. W. Larue. "Utilization of Existing Wells in Seismograph Work." *Early Geophysical Papers 12* (1931):119–127.

Quarles, M. "Vertical Seismic Profiling—A New Seismic Exploration Technique." Paper presented at the 48th Annals of International Meeting of SEG, (1978).

Rice, R. B., et al. "Developments in Exploration Geophysics, 1975–1980." *Geophysics 46* (1981):1088–1099.

Riggs, E. D. "Seismic Wave Types in a Borehole." *Geophysics 20* (1955):53–60.

Van Sandt, D. R. and F. K. Levin. "A Study of Cased and Open Holes for Deep Seismic Detection." *Geophysics 28* (1963):8–13.

Walton, G. G. "Three-Dimensional Seismic Method" Geophysics *37*, (1972):417–430.

Water, K. H. *Reflection Seismology, A Tool for Energy Resource Exploration* John Wiley and Sons, New York, (1987):538pp.

Wyatt, S. B. "The Propagation of Elastic Waves Along a Fluid-Filled Annular Region." Master of Science Thesis. University of Tulsa, Tulsa, OK, (1979).

Zimmermann, L. J. and S. T. Chen. "Comparison of Vertical Seismic Profiling Techniques." *Geophysics 58* (1993):134–140.

Chapter 5
Seismic Data Processing

Introduction

Data processing converts field recordings into meaningful seismic sections that reveal and help delineate the subsurface stratigraphy and structure that may bear fossil hydrocarbons.

The final interpretation of the seismic data is only as good as the validity of the processed data. It is imperative that the interpreter be aware of all the problems encountered in the field data acquisition and the data processing stage.

A data processing geophysicist must know and understand the regional geology of the Basin and particulars of each processing step. There is no a cookbook routine to follow in the processing. Each geologic setting presents its own specific problems to solve. Before routine processing for a prospect, extensive testing on the data should be done to study the problems involved to design the optimum parameters for each step in the data processing data flow.

Before discussing the data processing flow, some basic ideas and mathematical concepts should be covered.

Mathematical Theory and Concepts

As with any scientific-based discipline, seismic exploration makes considerable use of mathematics. Much of seismic data processing is based on a branch of mathematics called Statistical Communication Theory. Details of that theory are beyond the scope of this text but some of the basic concepts and applications of it will be presented. One aspect of modern seismic exploration methods results in a simplification of mathematical tools required. Since seismic data are recorded as *sampled* data, integration reduces to summation and differentiation to subtraction.

M.R. Gadallah, R. Fisher, *Exploration Geophysics*,
DOI 10.1007/978-3-540-85160-8_5, © Springer-Verlag Berlin Heidelberg 2009

Sampled Data

Data were recorded in *analog* form up until the late 1960s and early 1970s. At first data were recorded optically (light on photographic paper). In the 1950s analog magnetic tape was used. In the first case seismic data were represented by a continuous "wiggly trace" on photographic paper. In the latter case, seismic data were represented by continuously varying magnetic intensity.

For the past 40 years or so, seismic data have been recorded digitally. Digital recording means that the incoming analog signal is measured at regular time intervals and these measurements, in millivolts (mv), are written on magnetic tape as binary numbers. Thus, the input continuous signal is output as a set of numbers or *time series*. If the frequency content of the signal is consistent with the sampling increment, the original analog signal can be correctly reconstructed. Figure 5.1 illustrates the concepts of sampling and reconstruction.

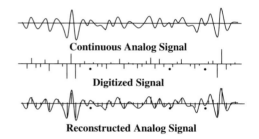

Fig. 5.1 Sampling and reconstruction of an analog signal

Figure 5.2 shows the effect of sampling at different *sample increments* or *sample periods*. Figure 5.2 shows four different analog inputs. All are single frequency signals at 50 Hz, 100 Hz, 200 Hz, and 300 Hz. Outputs (reconstructed analog signals)

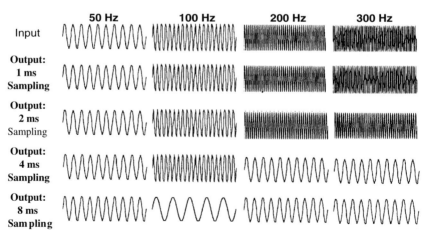

Fig. 5.2 Effect of frequency on reconstruction fidelity

sampled at 1 ms, 2 ms, 4 ms and 8 ms are shown for each input. In the case of the 50 Hz input, outputs are the same as the input at all sample periods. In the case of the 100 Hz sinusoid, the outputs sampled at 1 ms, 2 ms and 4 ms are the same as the input but the output sampled at 8 ms is a 25 Hz sinusoid! For the 200 Hz input the outputs for 1 ms and 2 ms sampling are the same as input (200 Hz). However, for 4 ms and 8 ms sampling the outputs are both 50 Hz. For the 300 Hz input only the output from 1 ms sampling is 300 Hz At 2 ms sampling the output is 200 Hz At 4 ms and 8 ms sampling the outputs are again 50 HZ.

What is going on here is that some of the inputs have been *aliased*. For every sample period there is a maximum frequency, called the *Nyquist frequency* beyond which data cannot be correctly reconstructed but, instead are aliased or folded over. The relationship between sample period and Nyquist frequency is listed in Table 5.1.

Table 5.1 Nyquist frequencies

Sample period in ms	Nyquist frequency in Hz
1	500
2	250
4	125
8	62.5

Table 5.1 can be used to analyze Fig. 5.2.

- Since none of the inputs are above 500 Hz, none are aliased at 1 ms sampling.
- At 2 ms sampling only the 300 Hz input is 50 Hz above the Nyquist frequency of 250 Hz and it aliases to 200 Hz, or 50 Hz below the Nyquist frequency.
- At 4 ms sampling both the 200 Hz and 300 Hz inputs are both above Nyquist. The 200 Hz input is 75 Hz above Nyquist so it aliases to 75 Hz below Nyquist or 50 Hz. The 300 Hz input is 50 Hz more than twice the Nyquist frequency so it also aliases to 50 Hz.
- At 8 ms sampling only the 50 Hz input is below the Nyquist frequency. The 100 Hz input is 37.5 Hz above the Nyquist frequency so it aliases to 37.5 Hz below the Nyquist frequency or 25 Hz. The 200 Hz input is 50 Hz below four times the Nyquist and it aliases to 50 Hz. The 300 Hz input is 50 Hz above four times the Nyquist and it also aliases to 50 Hz.

Aliasing is a type of distortion that must be avoided. Once it occurs, it cannot be removed,

Time (T), Frequency (F), Space (X) and Wavenumber (K) Domains

An extremely important concept in seismic data processing is that signal can be described equally well as a time series (amplitudes versus time) or as the combination of an *amplitude spectrum* (amplitudes versus frequency) and a *phase spectrum*

Fig. 5.3 Definition of phase:
top = zero-phase, *middle* =
60° phase lead, *bottom* = 45°
phase lag

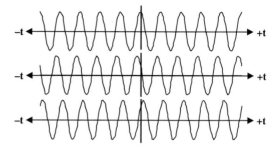

(phase versus frequency). In the first case the signal is defined in the time domain. In the second case the signal is defined in the frequency domain.

This, of course, brings up the question of what is meant by phase. A good beginning is to define zero-phase. A time series or function that is symmetrical about zero time is zero-phase. A cosine function is the prototypical example. Figure 5.3 shows a cosine wave centered at time zero and two time-shifted cosine waves. The cosine wave that is shifted to the left (toward negative or earlier time is said to have a *phase lead*. In this particular example the shift is 1/6 of a period or 60°. The cosine wave shifted to the right or toward later time is said to have a *phase lag,* here 1/8 of a period or 45°. Another way of stating this is to say that the three cosine waves of phases of 0°, +60°, and −45°.

Figures 5.4 and 5.5 provide another view of the relationship between the time and frequency domains. Figure 5.4(a) is the time domain description (a time series

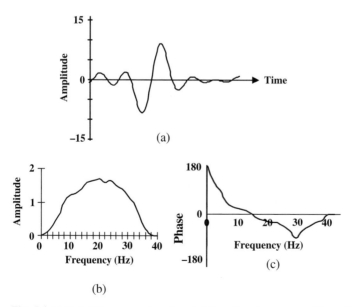

Fig. 5.4 Time and frequency domains; (**a**) time domain wavelet, (**b**) amplitude spectrum, and (**c**) phase spectrum

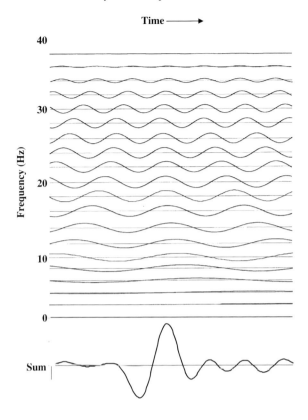

Fig. 5.5 Time domain wavelet as the sum of single frequency sinusoids

plot) while Fig. 5.4(b, c) are the frequency domain descriptions (amplitude and phase spectra, respectively). Figure 5.5 shows a set of single frequency sinusoids and their sum. The amplitudes of the sinusoids are given by the amplitude spectrum of Fig. 5.4(b) and the phases are given by the phase spectrum of Fig. 5.4(c). The sum in Fig. 5.5 is identical to the time domain wavelet of Fig. 5.4(a)!

It can be seen, then, that a time domain wavelet can be synthesized by summing a set of single frequency sinusoids. Conversely, a time domain wavelet can be decomposed into a set of single frequency sinusoids. The amplitude and phase spectra specify the component sinusoids so it is not necessary to show the individual sinusoids to show this equivalence of the time and frequency domains. The time domain wavelet and the combined amplitude and phase spectra are simply two ways of the describing the same thing.

To go from the time domain to the frequency domain requires use of the *Fourier Transform* and to go from the frequency domain to the time domain requires use of the *Inverse Fourier Transform*. For sampled, or digital, data simpler and faster methods called the *Fast* or *Finite Fourier Transform (FFT)* and the *Inverse Finite Fourier Transform (IFFT)* is used. The mathematics of these is beyond the scope of this book.

Just as a seismic signal can be defined in terms of change in amplitude with time at a particular location so can a seismic signal can be defined in terms of change in amplitude with distance, at a particular time – this is the space or X-domain. Also, note that seismic data are sampled in the space domain as well as in the time domain, since data are recorded only at discrete points. Thus, aliasing occurs in the K-Domain as well as in the F-domain.

In the time domain a particular signal component can be described by its *period (T)*, the time required to complete one cycle. As shown above, the frequency domain description is an equivalent way to describe seismic data and signal components are defined by frequency (F) or number of cycles/second. In the space domain a particular signal component can be described by its *wavelength (λ)*, the distance required to complete one cycle. In a similar way the wavenumber domain description is equivalent to that of the space domain and signal components are defined by wavenumber (k) or number of cycles/unit distance, meters or feet. As in the frequency domain, the wavenumber domain requires both an amplitude spectrum and a phase spectrum to completely define the data.

A particularly useful tool in seismic data processing is the transformation from time and space (T-X) domains to the frequency and wavenumber (F-K) domains. This is accomplished by applying a two-dimensional FFT. A two-dimensional IFFT is used to return from F-K to T-X.

Figure 5.6 illustrates T-X to F-K transformation by means of a schematic record. The direct arrival, first break refraction, and source-generated noise are linear in nature (i.e. – show a single dip) in both T-X and F-K. Signal (reflections) are non-linear and have many dips. Thus, in the F-K domain signal tends to lie in an area near K = 0.

The F-K transformed data are displayed in an area between $+k_N$ and $–k_N$, and from 0 to f_N. where k_N is the Nyquist wavenumber and f_N is the Nyquist frequency. The lines representing the direct arrival, first break refraction, and source-generated

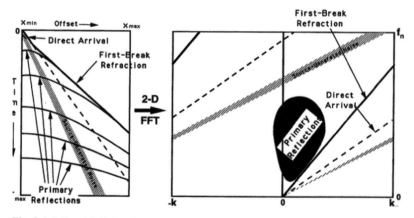

Fig. 5.6 T-X and F-K domains

noise are continued from the right edge of the F-K plane to the left edge because they have been aliased.

Note that signal is crossed by noise in the T-X plane but they are separated in the F-K plane. This provides a way of discriminating between signal and noise – more on this later in the chapter.

Filtering, Convolution, and Correlation

The surface of the earth is in constant motion because of both natural phenomena and cultural activities. This results in a background of earth motion that is recorded along with the desired motion produced by the seismic sources. This undesired recording is called noise. Thus, the amplitude spectrum of a seismic trace is the sum of signal and noise. The ratio of signal amplitude to noise amplitude is called the *signal-to-noise ratio (S/N)*.

It is desirable to have $S/N \geq 1$. This is usually the case over some range of frequencies but not all. In most cases $S/N < 1$ at the lowest frequencies and again at the highest frequencies. One way of accomplishing the desired result to apply a filter that greatly attenuates or, preferably zeroes, frequency components where $S/N < 1$. Such a filter is called a *band-pass filter*.

Figure 5.7(a) shows the amplitude spectrum of an ideal band-pass filter. It has a value of 1 for frequencies of 10 Hz through 32 Hz and 0 for all other frequencies Fig. 5.7(b) is the amplitude spectrum of Fig. 5.4(b). Multiplication of these two amplitude spectra produces the amplitude spectrum of Fig. 5.7(c), the filtered signal.

Figure 5.8 shows the equivalent operation in the time domain. Figure 5.8(a) is the inverse Fourier transform of the filter amplitude and phase spectra (called the

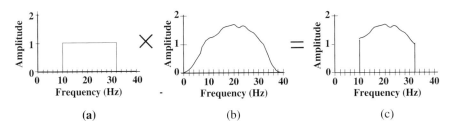

Fig. 5.7 Filtering in the frequency domain, (**a**) Filter amplitude spectrum, (**b**) input amplitude spectrum, and (**c**) output amplitude spectrum

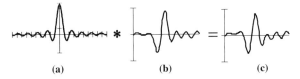

Fig. 5.8 Convolution or time-domain filtering; (**a**) filter impulse response, (**b**) input wavelet, (**c**) output wavelet

filter *impulse response*). Figure 5.8(b) is the wavelet of Fig. 5.4(a) and Fig. 5.8(c) is the *convolution* of the filter impulse response and the input wavelet. Note that an asterisk (*) indicates convolution.

Figure 5.8(a) was described as the filter impulse response but an impulse response was not defined. To determine an impulse response input a unit impulse (spike or Dirac delta function) into a system and record the system output. Figure 5.9 illustrates this by considering an example of utmost importance to seismic exploration – the earth's impulse response. Here the earth is considered to be a system. The response is a reflection from each reflecting horizon. The reflection signals are impulses with amplitudes equal to vertical reflection coefficients of the horizons and a time delays equal to their two-way reflection times.

Figure 5.10 shows two time series, A and B, and the result of their convolution, time series C. Convolution is NOT simple multiplication of one time series by another but it does involve multiplication. Table 5.2 illustrates a method for convolving two time series. The non-zero values of time series A are listed in row 1 of the table and the non-zero values of time series B are listed in column 1. To complete row 2 multiply the A values by 3. To complete row 3 multiply the A values by 1. To complete row 4, multiply the A values by −2. Sum these products diagonally as indicated by the arrows in Table 5.2 to obtain 9, 0, −7, and 2 – the values of the convolution as shown in Fig. 5.10.

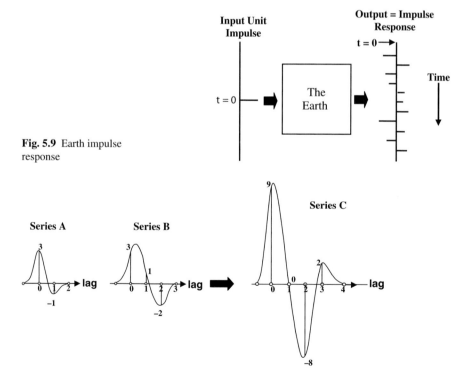

Fig. 5.9 Earth impulse response

Fig. 5.10 Convolution – Series A * Series B = Series C

Table 5.2 Matrix for convolution

B \ A	3	−1
3	9	−3
1	3	−1
−2	−6	2

Table 5.3 Matrix for crosscorrelation of series A with B

B \ A	3	−1
3	9	−3
1	3	−1
−2	−6	2

Crosscorrelation measures the similarity of one time series to another. With regard to mathematical computation, correlation and convolution appear similar. The same matrix as in Table 5.2 can be used to calculate the values of Series A with Series B. As shown in Table 5.3, the direction of summation is different. Figure 5.11 shows the result of crosscorrelating Series A with Series B.

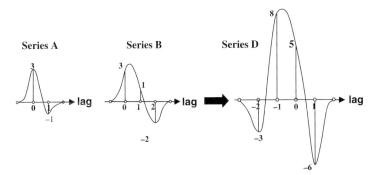

Fig. 5.11 Series A crosscorrelated with Series B = Series D

The convolution of Series A with Series B is the same as the convolution of Series B with Series A. The order of convolution makes no difference. In crosscorrelation it does make difference. The calculation is done by interchanging rows and columns in the matrix of Table 5.3, as shown in Table 5.4. The crosscorrelation of Series B with Series A is illustrated in Fig. 5.12. Comparing the two crosscorrelations shows that one is the time-reversed version of the other.

Table 5.4 Matrix for crosscorrelation of series B with A

A \ B	3	1	−2
3	9	3	−6
−1	−3	−1	2

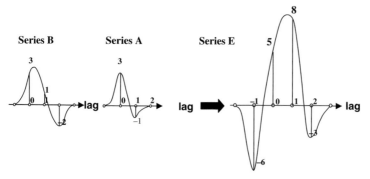

Fig. 5.12 Series B crosscorrelated with Series A = Series E

Crosscorrelation is when two different time series are correlated. Autocorrelation is the correlation of a time series with itself. It measures power of the time series. The autocorrelation is a symmetrical, zero-phase function, while a cross-correlation is not necessarily symmetrical.

Figure 5.13 shows the autocorrelations of Series A and Series B. Neither resembles their cross-correlation. Table 5.5 shows the calculation matrices for the two autocorrelations.

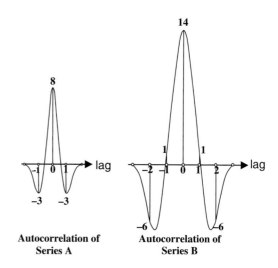

Fig. 5.13 Autocorrelations of   **Autocorrelation of**   **Autocorrelation of**
Series A and Series B   **Series A**   **Series B**

Table 5.5 Matrix for autocorrelations of series A and B

A\A	3	−1	B\B	3	1	−2
3	9	−3	3	9	3	−6
−1	−3	−1	1	3	1	−2
			−2	−6	−2	4

Processing Data Flow

Processing requires an orderly approach to converting raw field records into mean-ingful information about subsurface geology. The major steps in a processing seis-mic data include:

- Data initialization.
- Amplitude processing
- Pre-stack analysis
- Deconvolution
- Velocity analysis
- Statics Analysis (land and OBC)
- CMP Stack
- Migration
- Band-pass filtering

Data Initialization

In the early days of digital recording, field data wee recorded in multiplexed form, i.e. –ordered by sample time. (See Chap. 4.) To be able to process these data they had to be put in trace or channel order. This process is called *demultiplexing*. To-day, however, this should only be necessary when reprocessing very old data since modern seismic recording systems supply field data in demultiplexed form.

Normal processing in this stage includes:

- Convert data from field format (one of the standard SEG formats) to the format specified by the processing system being used
- Display all shot records to make sure documentation provided by the field crew is accurate and to help identify problems in the data
- Establish a *geometry database*.

    o Adopt an x-y coordinate system
    o Assign unique numbers to and specify coordinates for all source and receiver positions.

The geometry described by field documentation (shotpoint and spread locations) is the starting point for the geometry database. Every effort must be made to assure that it is correct and, if not, make corrections. Velocity and statics errors result from geometry errors. These errors cannot be removed except by correcting geometry errors.

To this point data are in field or *common shot* format but for parameter testing and selection a sort into *common midpoint (CMP)* format is needed. In processing land data, datum statics are calculated and applied in this stage. Calculations for refraction statics, if used, may be performed in this stage and saved for later use. Data required for both datum and refraction statics should be present in the updated geometry database.

Amplitude Recovery and Gain

Since it is known that lateral variation in the amplitudes of the reflections carry useful information, it becomes common practice to avoid tampering with the recorded amplitude values. However, it is critical to correct the amplitudes for the gradual decrease with depth due to spherical spreading and absorption.

Figure 5.14 is a comparison between field records before and after correcting the amplitudes for geometric spreading it can be seen that the amplitude is restored at late times on the record, but unfortunately, ambient noise has also been strengthened.

Fig. 5.14 Marine field record
(**a**) before and (**b**) after
geometric spreading
correction

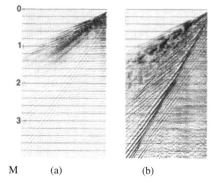

M   (a)   (b)

Types of Gain

Amplitudes are still not corrected for inelastic attenuation after correction for geometric spreading. The corrections for this are generally called *gain*. A great many types of gain have been developed and used. Among the more often used are:

- Programmed gain control (PGC)
- Automatic gain control (AGC)

  - RMS AGC
  - Instantaneous AGC

- Exponential gain

Figure 5.15 illustrates PGC. A set of multiplier, time pairs is selected by trial and error. The PGC function is developed by linear interpolation of multiplier values between these pairs. A constant multiplier equal to the first multiplier is used for times earlier than the first time selected. From the last time selected to the end of the record, the last multiplier selected becomes a constant multiplier.

AGC functions are determined statistically. Traces are divided into *time gates* and within these either the *average absolute* value or the root-mean-square (RMS) value amplitude is obtained. Figure 5.16 illustrates calculation of these two amplitudes in a gate. The AGC gain function is a set of multipliers for each sample time. In instantaneous AGC the multipliers are inversely proportional to average absolute values

Fig. 5.15 Programmed gain
control

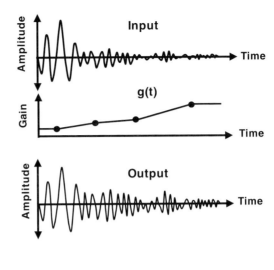

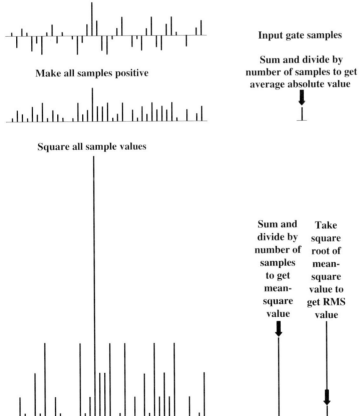

Fig. 5.16 Average absolute and RMS amplitudes in a single time gate

in each gate. In RMS AGC the multipliers are inversely proportional to RMS values in each gate. Since RMS amplitudes are usually larger than average absolute values amplitudes are usually changed more by RMS AGC than by instantaneous AGC.

Pre-Stack Analysis

In the case of marine data, pre-stack analysis includes selection of velocity analysis locations, identifying records and traces that require editing, and determining deconvolution parameters. In case of land data processing, especially for noisy data, pre-stack analysis is also done to analyze frequency content of the data and to choose parameters for processes that enhance signal-to noise ratio. Band-pass filter, velocity filter, and deconvolution tests are run for this purpose. Band-pass filtering discriminates between signal and noise on the basis of frequency. Velocity filters discriminate between signal and noise on the basis of apparent velocity. Decon can attenuate undesirable events such as short period multiples and enhance the vertical resolution, by collapsing wavelets. For marine data, designature may be run instead of (or in conjunction with) decon. Marine data normally have less random and coherent noise than land data, but they suffer from a higher degree of multiple reflections in many offshore areas. A common midpoint sort is generated for conducting many processing steps, such as applying elevation statics, velocity analysis, residual statics, and stack to name few. Figure 5.17 shows the flow chart of pre-stack analysis.

Distances and angle changes on the seismic line must be taken in consideration to obtain the correct distance from the source to receiver for each trace. Field statics (that account for trace to trace elevations differences) are usually applied before NMO to derive velocity analysis from seismic data.

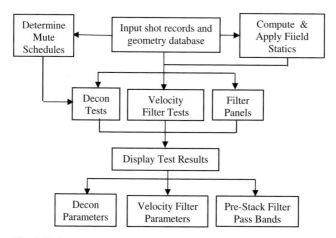

Fig. 5.17 Pre-stack analysis

There is no standard or routine processing sequence that will fit every seismic survey acquired in every part of the world, but there is an outline that must be followed. The processing sequence must be designed according to the area problems and record quality.

Mute

This is the process of excluding parts of the traces that contain only noise or more noise than signal. The two types of mute used are *front-end mute* and *surgical mute.*

Front-End Mute

In modern seismic work, the far geophone groups are quite distant from the energy source. On the traces from these receivers, refractions may cross and interfere with reflection information from shallow reflectors. However, the nearer traces are not so affected. When the data are stacked, the far traces are muted (zeroed) down to a time at which reflections are free of refractions. The mute schedule is a set of time, trace pairs that define the end of the muting.

Mute changes the relative contribution of the components of the stack as a function of record time. In the early part of the record, the long offset may be muted from the stack because the first arrivals are disturbed by refraction arrivals, or because of the change in their frequency content after applying normal moveout

The transition where the long offsets begin to contribute may be either gradual or abrupt. However, an abrupt change may introduce frequencies that will distort the design of the deconvolution operator. Figure 5.18 illustrates the front-end mute process.

Surgical Mute

Muting may be over a certain time interval to keep ground roll, airwave, or noise patterns out of the stack. This is especially applicable if the noise patterns are in the same frequency range as the desired signal. Convolution to filter out the noise may also attenuate the desired signal. Figure 5.19 illustrates the surgical mute approach.

Deconvolution Test

Deconvolution is discussed later in this chapter. The deconvolution operator or filter parameters include:

- Filter length
- Gap length
- Number of filters per trace
- Percent white noise to add

Fig. 5.18 Front-end mute.
(**a**) raw record with mute
defined. (**b**) record after front-
end mute

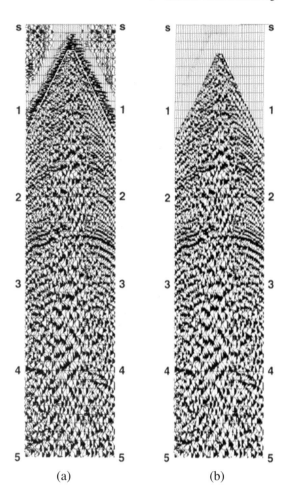

(a)   (b)

Deconvolution testing consists of trying different values of these parameters in the deconvolution of selected CMP gathers. The values that yield the best result are chosen for deconvolution application. It is very important that only one parameter be changed at a time.

Velocity Filter Tests

Velocity filtering is used to attenuate source-generated noise that travels along or parallel to the surface. Such noise has a single constant velocity or propagates as a set of noise trains, each having a constant velocity. Testing usually involves transforming selected shot records from the T-X domain into the F-K domain. Figure 5.20 shows such a record in the F-K domain. A triangular area where noise dominates is shown. The straight lines drawn correspond to particular velocities. The parameters selected are pairs of velocities defined by these lines.

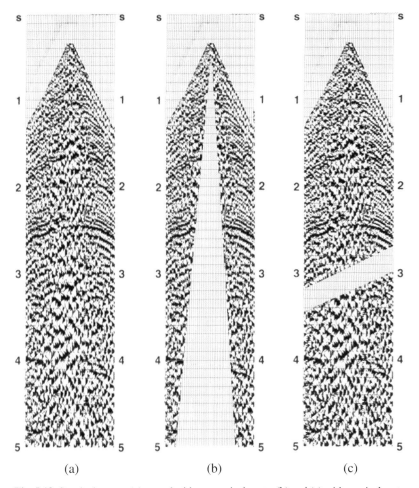

Fig. 5.19 Surgical mutes. (**a**) record without surgical mute. (**b**) and (**c**) with surgical mutes applied

Filter Test

One application of convolution process is to apply band-pass filter to seismic data to attenuate some undesired frequency range. This can be done by obtaining a display of a seismic record that is filtered with different filter pass bands. Figure 5.21 shows an example of filter scans (right) on raw field data (left). the band pass of the filter used for each panel is annotated at the head of each panel. The low-frequency signals are at least 10 Hz (and probably lower). The high-cut of the filter should be time-variant here: higher than 60–70 Hz could be cut after 1.5 s; higher than 50–60 Hz could be cut after 1.5 s; higher than 50–60 Hz could be cut after 2.4 s; higher than 40–50 Hz could be cut after 4 s.

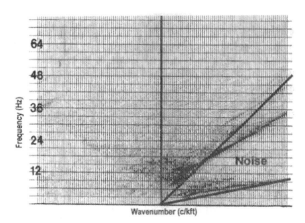

Fig. 5.20 2-D fourier transform of a shot record

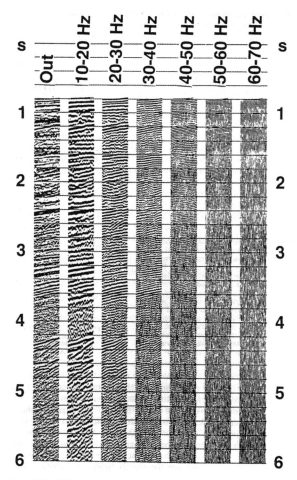

Fig. 5.21 Filter scans

The filters designed from analysis of the filter scans are applied to the data after each processing step or stage to better evaluate effect of parameters used. Outputs from one processing stage to the next should not be filtered.

Static Corrections

Two main types of correction need to be applied to reflection times on individual seismic traces in order that the resultant seismic section gives a true representation of geological structure. These are *dynamic* and *static corrections*. The static correction is so-called because it is a fixed time correction applied to the entire trace. A dynamic correction varies as a function of time as in application of NMO corrections.

In order to obtain a seismic section that accurately shows the subsurface structure, the datum plane at a known elevation above mean sea level and below the base of the variable-velocity weathered layer (see Fig. 5.22). The value of the total statics (ΔT) depends on the following factors:

1. The perpendicular distance from the source to the datum plane.
2. Surface topography; that is, the perpendicular distance from the geophone to the datum plane.
3. Velocity variations in the surface layer along the seismic line.
4. Irregularities in thickness of the near-surface layer.

In computing ΔT it is usually assumed that the reflection ray path in the vicinity of the surface is vertical.

The total correction is:

$$\Delta T = \Delta t_s + \Delta t_r$$

where

$$\Delta t_s = \text{the source correction, in ms}$$
$$\Delta t_r = s\text{the receiver correction, in ms}$$

As indicated in Fig. 5.22, elevation or datum statics effectively move source and receiver from the surface to the datum plane.

Figure 5.23 shows common shot gathers from a seismic land line, where statics (due to near-surface formation irregularities) caused the departure from hyperbolic travel times on the gathers at the right side of the display.

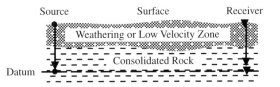

Fig. 5.22 Static corrections

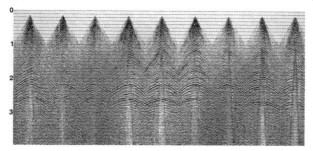

Fig. 5.23 Shot records with statics problems

Deconvolution (Decon)

Vertical resolution is defined as how closely two seismic events can be positioned vertically, yet be identified as two separate events. Vertical resolution, or *tuning thickness,* is usually considered to be equal to one-fourth the dominant wavelength, which is a function of dominant frequency and velocity of the target horizon. If the signal-to-noise ratio is greater than 1:1, vertical resolution is improved by improving the bandwidth.

Deconvolution is a process that improves the vertical resolution of seismic data by compressing the basic wavelet, which also increases bandwidth of the wavelet. In addition to compressing or shortening reflection wavelets deconvolution can also be used to attenuate ghosts, instrument effects, reverberations and multiple reflections. The earth is composed of layers of rocks with different lithologies and physical properties. In seismic exploration, their densities and the velocities at which the seismic waves propagate through them define rock layers. The product of density and velocity is called acoustic impedance. It is the impedance contrast between layers that causes the reflections that are recorded along a surface profile.

A seismic trace can modeled as the convolution of the input signature with the reflectivity function of the earth impulse response, including source signature, recording filter, surface reflections, and geophone response. It is also has primary reflections (reflectivity series), multiples, and all types of noise. If decon were completely successful in compressing the wavelet components and attenuating multiples it would leave only the reflectivity of the earth on the seismic trace. In so doing, vertical resolution is increased and earth impulse response or reflectivity is approximately recovered. This is demonstrated in Fig. 5.24. Decon is normally applied before stack (DBS). But it is sometimes is applied after stack (DAS) The term deconvolution is most often applied to a type of inverse filtering. Figure 5.25 defines inverse filtering. A wavelet or time series is input (convolved with) the inverse filter and a unit spike or impulse is output. The problem with this approach is that the inverse filter has to be infinitely long, which of course is not possible. Instead of an infinitely long filter, one of a practical length that will produce the desired output with the least squared error is designed. Figure 5.26 is an example. The input time series is $\{4, -3, 2\}$ and the desired output time series is $\{1, 0, 0, 0, 0\}$. The least

Fig. 5.24 Deconvolution
(DECON) objective

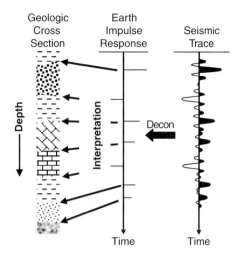

Fig. 5.25 Inverse filter
definition

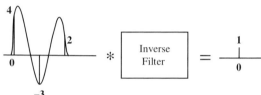

square error approximation filter is $\{0.232, 0.168, \text{and } 0.040\}$. Convolving the filter
with the input yields an output of $(0.928, -0.024, -0.012, 0.048, 0.040\}$. This is not
exactly the desired output but quite close and only using three filter points!

This type of inverse filtering is also called spiking or whitening decon. The left
side of Fig. 5.27 is a relative amplitude spectrum similar to what might be expected
for a trace from a record before decon. The center of Fig. 5.27 is the amplitude
spectrum of the infinite length inverse filter (simply the reciprocal of the input amplitude spectrum). The product of these two amplitude spectra is on the right side of
Fig. 5.27 – a flat spectrum (constant value at all frequencies). This is called a *white
spectrum* because it is analogous to the spectrum of white light, which is contains
equal amounts of all frequencies of light.

Since infinite length filters are not realizable, approximations must be used as
indicated above. The outputs of such filters have broader and smoother amplitude
spectra than the inputs but do not completely whiten the spectra. See Fig. 5.28.

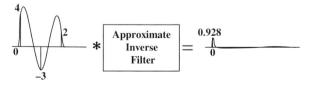

Fig. 5.26 Least square error approximate inverse filter

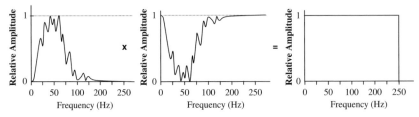

Fig. 5.27 Inverse filtering in the frequency domain

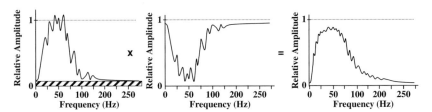

Fig. 5.28 Whitening decon in the frequency domain

Sometimes decon filter design "blows up" because a divide by zero is attempted. To avoid, or correct for this, "white noise" is added to the input. Adding white noise is adding a constant amount to the amplitude spectrum of the input as indicated by cross-hatched area at the bottom of the input amplitude spectrum (left side of Fig. 5.27).

Since decon filters are usually designed in the time domain, using the trace auto-correlation, a convenient, and equivalent, alternative to this addition of "white noise" is to increase the zero-lag value of the autocorrelation by a small percentage of it.

Adding white noise, or "pre-whitening", should be done only when necessary to get a filter designed. Only just enough white noise to achieve filter design should be added. Sometimes collapsing a wavelet is not desirable early in the processing flow. However, it is desirable to remove or attenuate the effects of ghosts, multiple reflections, instrument response, etc.. In this case the usual action is to design and apply *a prediction error filter* or *gapped decon*.

The desired output of the prediction error filter is the input values at some time in the future; say (n) samples ahead. There are two steps in filter design. First is the prediction filter. In Fig. 5.29 the desired output is the time series $\{4, -3, 0, 0, 0, 0, 0\}$. That is, the wavelet is to be shortened by one sample. This means that the value 2, one sample ahead, must be predicted. The 3-point filter that best predicts this is $\{0.463, 0.339, 0.082)$. To achieve the second step the prediction error filter is developed. In the case shown in Fig. 5.28 this is 11, 0, $-0.463, -0.339, -0.082$). The initial value of 1 establishes the point from which the prediction is made. The single 0 indicates that the prediction is to be made is one sample period. (This is the *gap*.) The remaining values are the prediction filter with their algebraic signs reversed. The sign reversal means that the predicted values are subtracted in the convolution process. The shape of reflection wavelets observed on seismic traces is the result

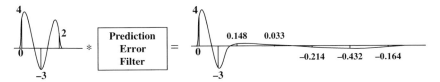

Fig. 5.29 Prediction error filter

of several effects that occur at the source, transmission through the earth, and in the recording process. These effects include the source wavelet, the recording instrument filter responses, the ghosting function, the reverberation operator, and the multiple operators. Usually none of these are known explicitly. However, the trace autocorrelation function provides information such as source wavelet shape, periods of multiples and reverberations, and amplitude relationships between multiples and primaries as shown by the *autocorrelogram* of Fig. 5.30.

A decon filter should be somewhat longer than the period of the multiple it is to attenuate. The whitening decon filter includes approximate inverses of the source wavelet, the recording instrument filter responses, the ghosting function, the reverberation operator, and the multiple operators. The prediction error or gapped decon does not usually include an inverse filter for the source wavelet. To be exact inverses, as previously noted, the filters would have to be infinitely long.

The length of the gap (α) in a gapped decon filter determines the amount of wavelet shortening, or whitening, that the filter does. Many people feel that an optimum value for gap length is the time of the second zero crossing of the autocorrelogram function this parameter. See Fig. 5.30.

Figure 5.31 shows the effect of deconvolution by comparing records with, and without, decon applied. Note that the reflection widths are decreased and a number of events are removed or greatly attenuated in the deconvolved record compared to the one not deconvolved. These events are likely multiple reflections.

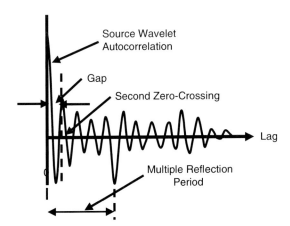

Fig. 5.30 Trace autocorrelogram and information it contains

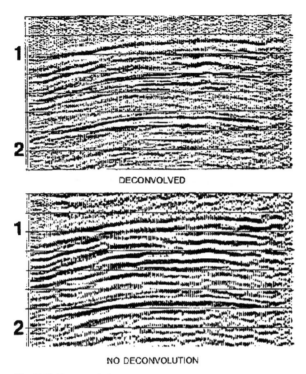

Fig. 5.31 Deconvolution versus No deconvolution

Because of inelastic attenuation, reflection wavelet shapes change with record time. Since the later reflections propagate over greater distances, the wavelets are subjected to more attenuation of the higher frequency components. Figure 5.32 illustrates how wavelets become progressively lower frequency and longer in time duration as record time increases.

As a result of this change in wavelet shape, a decon filter designed from and applied to a whole trace will not be effective at all record times. To alleviate this problem, multiple filters are designed and applied over different time intervals of a seismic record. The filter outputs are smoothly merged together to give a single

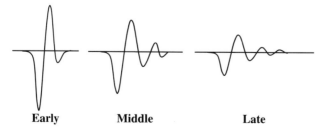

Fig. 5.32 Change in reflection wavelet shape with record time

time-variant deconvolution. For this reason the common reference to conventional trace Deconvolution is *time-variant deconvolution (TVD)*.

Various other approaches to the general deconvolution problem are in use. Among these are:

- *Signature deconvolution*
- *Surface-consistent deconvolution*
- *Time-variant Spectral Whitening (TVSW)*
- *Model-based Wavelet Processing (MBWP)*

Signature deconvolution is a type of deconvolution often applied to marine data. In marine operations, a hydrophone is suspended a 100 m or so below the airgun array and the energy detected by this hydrophone is recorded in a special channel of the recording instruments. Signature deconvolution designs a least square error approximate inverse of the recorded "signature" and applies this inverse filter to all traces recorded at this shotpoint. Note that signature decon does not affect changes in wavelet shape with record time caused by inelastic attenuation.

Surface consistent deconvolution uses the redundancy of multifold data to statistically determine and attenuate effects on signal waveform that occur in the vicinity of each source and receiver position.

Time-variant Spectral Whitening (TVSW) corrects for inelastic attenuation by dividing the input into relatively narrow frequency bandwidths and applying AGC to each. The separate frequency bands are then recombined in such a way as to preserve original amplitude levels. Note that TVSW can be effective in whitening the spectrum but has no effect on multiple reflections, ghosts, etc.

The parameter that describes the frequency-dependent attenuation effect of subsurface rocks is denoted as Q. Model-based wavelet processing develops mathematical models for the source wavelet, noise, and the Q-effect. It is a very effective tool in removing differences caused by different energy sources being used along a single seismic line.

Inverse-Q filters attempt to determine average values of Q over different time zones of a seismic section and use these values to design and apply filters in the frequency domain that compensate for inelastic attenuation.

Velocity Analysis

The word velocity seldom appears alone in seismic literature. Instead it will occur in combinations such as *instantaneous velocity, interval velocity, average velocity, RMS velocity, NMO velocity, stacking velocity, migration velocity, apparent velocity*, etc. Figure 5.33 can be used to illustrate some of these.

The velocity indicated by $V(x_a, z_a)$ is the velocity that would be measured at a point a distance x_a from the left of the Figure and at a depth z_a is an example of instantaneous velocity.

The velocities indicated as V_1, V_2, V_3, etc. are interval velocities. They are the average velocity through an interval of depth or record time and equal the thickness of the depth interval divided by vertical time through the interval.

Fig. 5.33 Velocity types

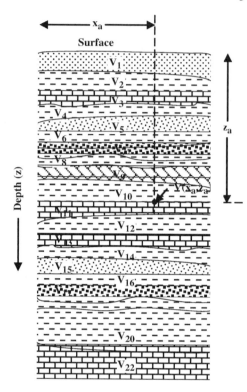

Average velocity to a particular depth is simply the depth divided by the time it takes a seismic wave to propagate vertically to that depth. Since seismic wave propagation times are usually measured as two-way times, the average velocity to, say, z_a, in Fig. 5.32 is $(2z_a/T_a)$, where T_a is the two-way time to depth z_a. Average velocity is required to convert time to depth.

Root Mean Square or *RMS* velocity at a particular record time, T_n, is calculated as follows:

1. Determine what interval times sum to the value T_n
2. Square the corresponding interval velocities
3. Multiply the squared interval velocities by their interval times
4. Sum the products obtained in step 3
5. Divide the sum obtained in step 4 by T_n
6. Take the square root of the value resulting from step 5, This is the RMS velocity at time T_n

If all reflectors are flat or nearly flat, RMS velocity is the same as NMO velocity.

NMO velocity is the velocity used to correct for Normal NMO. If NMO corrections are correct, and no other factors are involved, all primary reflections on CMP gather records occur at the same time on all traces.

Stacking velocity is the velocity that gives the optimum common midpoint (CMP) stack output when it is used for NMO corrections. It may be the same as NMO velocity but if there is significant dip on reflectors, it probably will not be the same.

Migration velocity is the velocity that optimizes the output of a migration algorithm, i.e. – moves the reflected energy to the correct times and places.

Apparent velocity is determined by dividing a horizontal distance by the time a seismic signal appears to propagate across it. For source-generated noise, apparent velocity and propagation velocity are equal but for reflections, it is much faster than NMO or stacking velocity. Apparent velocity is important in designing velocity or F-K filters. Stacking velocity, migration velocity and average velocity are the most important to seismic data processing.

Types of Velocity Analysis

Several methods of velocity analysis have been used over the years. The more commonly used methods are:

- $T^2 - X^2$ Analysis
- Constant Velocity Stack
- Velocity Spectrum Method

$T^2 - X^2$ **analysis.** In this method, times (T) of selected primary reflections on each trace are squared and plotted against the square of the offset (X) corresponding to the traces on which times were picked. Equation (3.6) of Chap. 3, repeated below, gives the relationship between T^2 and X^2. Based on this equation, the plotted points should fall along a straight line.

$$T_x^2 = \frac{x^2}{V^2} + T_0^2$$

Figure 5.34 illustrates a $T^2 - X^2$ plot for a particular primary reflection. Note that the plotted points do not fall exactly along the straight line that is fitted to them. Noise and residual statics tend to cause deviations of times.

The slope of the straight line is $1/V_{NMO}^2$ and the intercept at $X = 0$ is T_0^2. The least square fitting method is used to define line slope for all significant primary reflections. The $T^2 - X^2$ method is a reliable way to estimate stacking velocity.

Velocity Analysis Methods

Normal moveout is the basis for determining velocities from seismic data. Computed velocities can be used to correct for normal move out so that reflections are aligned on CMP traces. The accuracy depends on signal to noise ratio because it affects the quality of the reflection time picking.

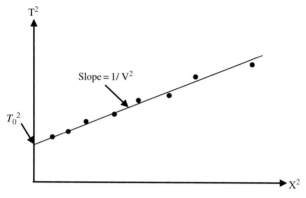

Fig. 5.34 $T^2 - X^2$ analysis

Constant Velocity Stacks (CVS)

Stacking velocities are often estimated from CMP gathers by evaluating stacked-event amplitude and continuity when a single velocity is used for NMO corrections at all times on the record. A range of velocities defined by the lowest and highest stacking velocities expected. A set of velocities is generated by incrementing by a constant amount from the lowest to the highest velocity.

Figure 5.35 shows a set of constant velocity stack panels. In this example, a portion of a line consisting of 24 common midpoint gathers has been NMO-corrected and stacked with velocities ranging from 5,000 to 13,600 ft/s. The resulting stacked 24 traces set, displayed as one panel, represent one constant velocity. The panels are

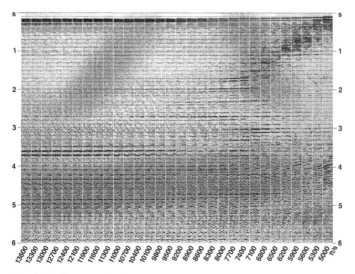

Fig. 5.35 Constant velocity stack display

displayed side by side with the velocity values indicated beneath each panel. Velocity values increase from right to left. Stacking velocities are picked directly from these panels by selecting the velocities that yield the best coherency and strongest amplitudes for velocity values at a certain center times. Notice that the deep events between 3.0 and 4.0 s seem to stack about the same over a wide of velocities. This represents the decrease of resolution of the stacking velocities with depth, caused by the decrease of NMO with depth (related to the increase of the velocity). The qualitative nature of this method makes it difficult to achieve the accuracy, or resolution, that is needed for good stacks.

Velocity Spectrum Method

Accurate and efficient velocity analysis requires the following:

- Restriction to velocity values that can be reasonably expected, in a time-variant manner
- Statistical evaluation of stack response (reflection amplitude and continuity) using small time gates and many estimates of stacking velocity
- Adequate sampling in time and velocity Displays of velocity calculation results that allow accurate velocity interpretation

The initial result of velocity analysis is a set of velocity functions that are determined at specific CMP locations within the survey. As shown in Fig. 5.36, velocity functions are defined by sets of time, velocity pairs that are picked for significant primary reflections. Linear interpolation between these points defines velocities for every sample in the CMP traces. Spatial interpolation between and extrapolation from these functions establishes a *velocity field* that defines velocities for every sample in the data volume.

The velocity spectrum approach is based on the correlation of the traces in a CMP gather. This method is suitable for data with multiple reflection problems, and less suitable for highly complex structure problems.

Figure 5.37 shows a model CMP record, without NMO corrections, on the left. The center and right demonstrate two ways of presenting the statistical evaluation of stack response. Both employ the statistical measure of *coherency*, that is related to crosscorrelation.

Figure 5.37 displays coherency values for five CMPs along a single line. Differences among them are evident. It is obvious that velocity functions derived by interpreting these displays will also be very different and spatial interpolation is necessary to adequately define the velocity field. The number of coherency values displayed for each CMP in Fig. 5.38 indicates the large number of small time intervals over which coherency is evaluated.

The velocity spectrum method distinguishes the signal along the hyperbola paths even with high random noise level. This is because of the power of cross-correlation in measuring coherency. The accuracy of the velocity spectrum is limited, however, if S/N ratio is poor.

Fig. 5.36 Velocity function

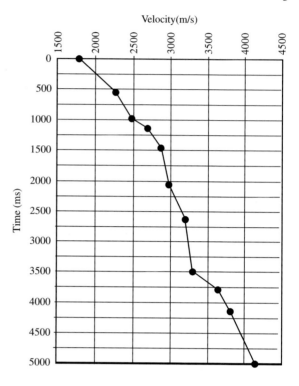

Fig. 5.37 Velocity spectrum statistical analysis

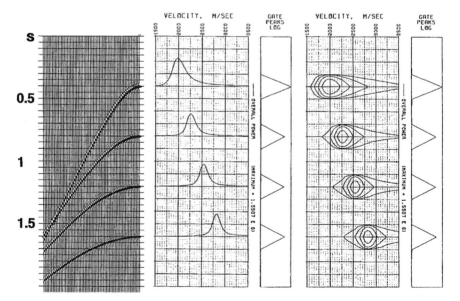

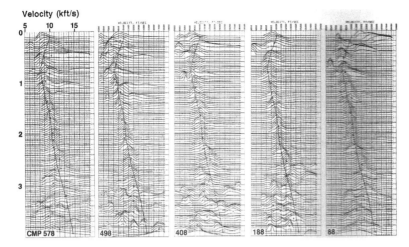

Fig. 5.38 Change in velocity spectra in space

Factors that Affect Velocity Estimates

There are several factors that affect the estimation of velocity using the common depth family of the seismic data:

- Field geometry- long offset
- Multiplicity- stacking fold
- Signal- to-noise ratio
- Front- end mute
- Time gate length for estimating the velocity estimates
- Velocity increment
- Coherency measures
- Data frequency range- Bandwidth
- Statics-true departure from hyperbolic normal move out.

Spread length is important factor in deriving good stacking velocity values. The longer the spread, as long as it does not exceed depth to target, the better the NMO values and therefore the better the discrimination between primary and multiple reflections. Depending on the record quality in an area, the fold can affect the signal-to-noise ratio, as the higher the multiplicity the better random noise attenuation. The details of the shallow events on the seismic section are as important as the target zone and can contain a very valuable information that affect the fidelity of a prospect. Therefore, front end muting is very crucial in eliminating noise trains such as body waves. Care must be taken in selecting the mute and protecting (not muting) as many short traces (close to the source) in order to enhance the coherency of the shallow reflections. If a coherent noise pattern, such as surface waves and airwaves, cannot be attenuated effectively by digital filter application (frequency of the noise is the same as a primary) a surgical mute can be used to eliminate the noise train and accordingly, obtain a better velocity analysis.

Resolution of the velocity estimates depends on the time gate length (normally 20 ms) over which the normal moveout calculations are made, the velocity increment and the velocity range. Both factors affect the resolution of the estimates and the run time on the computer. The frequency content of the data has a great effect of the velocity details especially shallow on the section. It affects the coherency and the lateral and vertical resolution.

Residual Statics Analysis

Applying NMO and elevation statics corrections to traces within the CMP gather may not result in perfect alignment of primary reflections. Misalignment may be caused by incorrect or incomplete static corrections. The presence of these "residual" statics generates poor stack traces. To correct these misalignments an estimate of the time shifts from perfect alignment is needed. There are two basic approaches to the problem. One is based on primary reflections and the other is based on first break refractions. Within the reflection-based approach are two methods: conventional and surface consistent.

Reflection-Based Residual Statics Analysis

Conventional reflection-based residual statics analysis is rather straightforward. It is assumed that the residual statics are random and, thus, that primary reflections stack to the correct time. (See Fig. 5.39.) If the individual traces in a CMP are crosscorrelated against a model trace derived from their stack, the time at which the crosscorrelation maximum occurs will be the residual static for that trace. See Fig. 5.40.

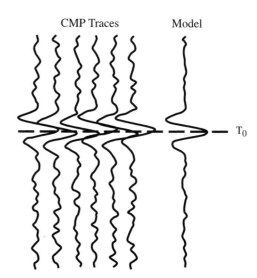

Fig. 5.39 Construction of model trace for crosscorrelation

Fig. 5.40 Determination of
Δt, static time deviation from
crosscorrelation with model
trace

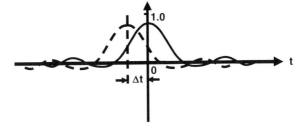

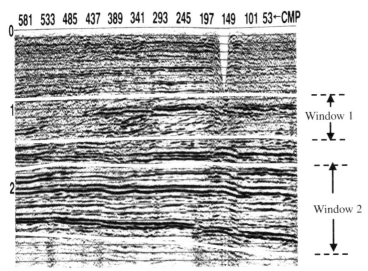

Fig. 5.41 Residual statics analysis windows

As shown in Fig. 5.41, one or two time windows are defined on a CMP stack. (The authors prefer two windows.) Note that the windows follow apparent structure on the stack section. Only the trace segments defined by the windows are used in building the trace models. Residual statics defined by the crosscorrelations are saved for application in the next processing step.

Surface-Consistent Residual Statics

A model is needed for the moveout-corrected travel-time from a source location to a point on the reflecting horizon, then back to a receiver location (sees Fig. 5.42). The key assumption is that the residual statics are surface consistent, meaning that static shifts are time delays that depend on the sources and receivers on the surface (Hileman et al., 1968; Taner et al., 1974). This assumption is valid if all raypaths, regardless of source-receiver offset, are vertical in the near-surface layers.

Since the near-surface weathered layer has a low velocity value, and refraction in its base tends to make the travel path vertical, the surface-consistent assumption

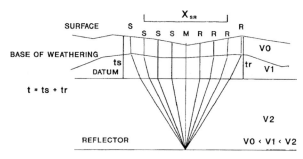

Fig. 5.42 Propagation model for surface-consistent statics

usually is valid. However, this assumption may not be valid for high-velocity per-mafrost layers in which rays tend to bend away from the vertical.

Residual static corrections involve three stages:

1. Picking the values.
2. Decomposition of its components, source and receiver static, structural and nor-mal moveout terms.
3. Application of derived source and receiver terms to travel times on the pre-NMO corrected gathers after finding the best solution of residual static corrections. These statics are applied to the deconvolved and sorted data, and the velocity analysis is re- run. A refined velocity analysis can be obtained to produce the best coherent stack section.

Figure 5.43 is a recommended flow chart for residual static correction and veloc-ity analysis. Figure 5.44 shows the effect of residual statics application on velocity analysis and velocity picks. Figure 5.45 shows a group of CMP families with nor-mal moveout applied. The top section, is before residual static application, Bottom section is after application of the static correction to each individual trace within

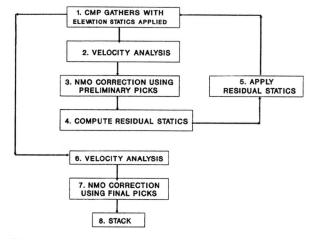

Fig. 5.43 Flow chart for surface-consistent statics analysis

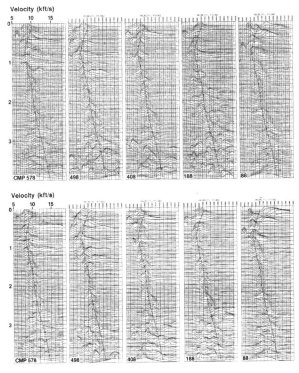

Fig. 5.44 Effect of residual statics on velocity picks (Reprinted from O. Yilmaz, Seismic Data Processing, 1987, Courtesy of Society of Exploration Geophysicists.)

the CMP family. One expects better coherency on the stack after residual static is applied. Figure 5.45a shows more severe static problems, and Fig. 5.45b shows the dramatic improvement by after application of the surface consistent residual static. Figure 5.45a is a CMP stack obtained applying elevation static for elevation changes. Figure 5.45b is the stack after applying surface-consistent static. Notice the great improvement at the right side of the section. Velocity analysis refinement can yield better velocity picks that result in a better stack.

New velocity picks are obtained by re-running the velocity analysis after applying residual statics to pre-NMO sort data. These generate a superior stack, as shown in the post-static version of the stack. See Fig. 5.46.

Refraction Statics

In the example of Fig. 5.45, surface-consistent residual statics removed the trace to trace time differences caused by near-surface variations, resulting in better continuity and coherency on the stack section. These trace to trace statics are called *short-period statics*.

An important question in estimating the shot and receiver static is the accuracy of the results as a function of the wavelength of static anomalies. Surface-consistent

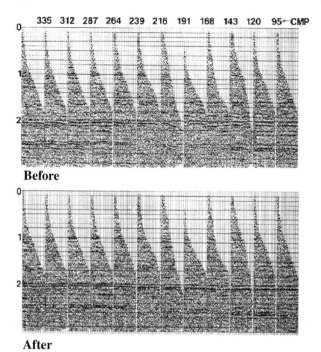

Fig. 5.45 NMO gathers before and after residual statics (Reprinted from O. Yilmaz, 1987, Seismic Data Processing, Courtesy of Society of Exploration Geophysicists.)

residual statics cannot solve long-period static problems. Figure 5.47 shows that the residual static application yields a much improved stack response. Short-period static shifts (less than one spread length) cause travel time distortion, which degrades the stack section. Correcting for short-period statics is not enough.

The structure at point A, between CMPs 149 and 197, is probably created by long-period statics. This can be identified by tracking the shallow horizon and the elevation profile. This example suggests that the elevation statics were not applied adequately. Residual static corrections are needed because the elevation or datum static corrections are not enough to compensate for irregularities in the near-surface formations. This is because the lateral variations of velocity in the weathered layer are unknown.

Residual static correction does a good job on the short-period static, but it does a poor job on the long-period static. The reason is that the residual static programs work on the arrival time difference between traces and not on absolute time values.

The refraction static method works on the absolute values of the first break arrival times. Figure 5.48(a) illustrates selected CMP gathers from a stack section. Figure 5.48(b) shows the gathers after applying linear moveout correction. A velocity value was chosen to flatten the refraction breaks. Figure 5.48(c) is the stack of the section after applying linear move-out to the gathers. Figure 5.48(d) is the stack after applying long-period statics. Statics were applied after NMO corrections and mute.

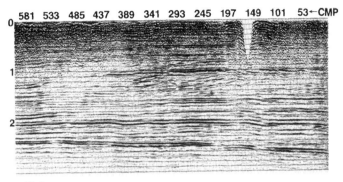

Before

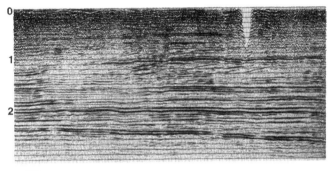

After

Fig. 5.46 CMP stacks before and after residual statics (Reprinted from O. Yilmaz, 1987, Seismic Data Processing, Courtesy of Society of Exploration Geophysicists.)

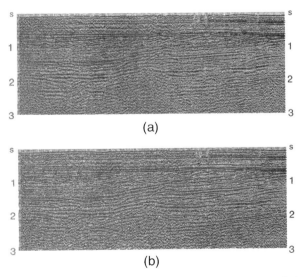

(a)

(b)

Fig. 5.47 (**a**) CMP stack with short period residual statics applied. (**b**) same CMP stack with both short and long period residual statics (Reprinted from O. Yilmaz, 1987, Seismic Data Processing, Courtesy of Society of Exploration Geophysicists)

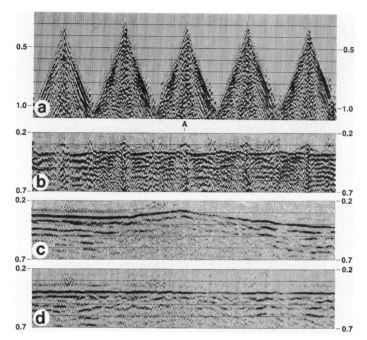

Fig. 5.48 Refraction statics method (Reprinted from O. Yilmaz, 1987, Seismic Data Processing, Courtesy of Society of Exploration Geophysicists.)

From the previous discussions, we can conclude that in case of severe weathering problems on land data statics may be applied as follows:

1. Elevation static corrections to account for elevation changes.
2. Refraction-based static corrections to remove long-period anomalies in a surface-consistent manner.
3. Conventional or surface-consistent refraction-based residual static connections to remove any remaining short-period static shifts.

Stacking

There is, actually, more than one kind of stacking. Vertical stacking, usually done by the acquisition crew, is performed to combine all vibrator records from a single source pattern into one record. Horizontal or, more commonly called, CMP stacking is the first output of seismic data processing that can be readily interpreted. It combines all traces of all CMP gathers into single traces for every CMP on a line. The common element is that two or more traces are combined into one.

This combination takes place in several ways. In digital data processing, the amplitudes of the traces are expressed as numbers, so stacking is accomplished by adding these numbers together. Peaks appearing at the same time on each of two traces combine to make a peak twice as high as one. The same is true of two troughs. A peak and a trough of the same amplitude at the same time cancel each other, and

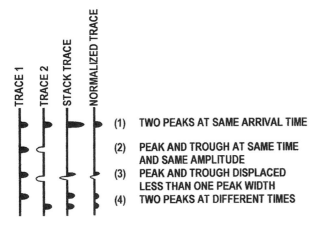

(1)   TWO PEAKS AT SAME ARRIVAL TIME

(2)   PEAK AND TROUGH AT SAME TIME
      AND SAME AMPLITUDE

(3)   PEAK AND TROUGH DISPLACED
      LESS THAN ONE PEAK WIDTH

(4)   TWO PEAKS AT DIFFERENT TIMES

Fig. 5.49 CMP stack principle

the stack trace shows no energy arrival at that time. If peaks are at different times, the combined trace will have separate peaks of the same sizes as the original ones. After stacking, the traces are "normalized" to reduce the amplitude so that the largest peaks can be plotted. Figure 5.49 illustrates the principle of stacking.

Applications of stacking include testing normal moveout, determining velocities, and *attenuating* noise to improve signal-to-noise ratio. The combination of six traces in a CMP family is a 600% stack. Similarly, the combination of 24 traces in a CMP family is a 2,400% stack.

Prior to the actual stack NMO and residual static corrections, if applicable, should be applied. The final CMP stack will use the velocity field obtained from the last velocity analysis run. Similarly, residual statics should be from the last run of residual statics. In addition, front end mutes (and surgical mutes, if needed) should be applied to remove distorted and/or unwanted energy.

Normal moveout (NMO) corrections are applied to correct for the horizontal component of reflection ray paths. If the recorded time at offset x is T_X, and the zero-offset time is T_o, a correction ΔT_X is required to be subtracted from T_X., i.e. $T_o = T_X - \Delta T_X$. The correction ΔT_X increases as offset increases and decreases as both velocity and record time increase.

Previously it was noted that static corrections, in effect, move sources and receivers from the surface to the datum. NMO corrections convert all times to zero offset times and CMP stack, in effect, moves all sources and receivers to their common midpoint positions.

Migration

A seismic section is assumed to represent a cross-section of the earth. The assumption works best when layers are-flat, and fairly well when they have gentle dips. With

steeper dip the assumption breaks down; the reflections are in the wrong places and
have the wrong dips.

In estimating the hydrocarbons in place, one of the variables is the areal extent of the trap. Whether the trap is structural or stratigraphic, the seismic section
should represent the earth model. *Dip migration,* or simply *migration,* is the process of moving the reflections to their proper places with their correct amount of
dips. This results in a section that more accurately represents a cross-section of the
earth, delineating subsurface details such as fault planes. Migration also collapses
diffraction.

Normal Incidence

Figure 5.50 (a) shows the simple model of a flat reflector. Energy from the source
goes straight down to the reflecting horizon and back up to a geophone at the shot
point. If the horizon dips, the energy goes to and from it by the most direct path,
which is along a normal to the reflector. Energy that strikes the reflector at other
angles goes off in another direction, as shown in Fig. 5.50(b). The normal incidence
principle is the basic idea behind dip migration. All post-stack migration techniques
follow this principle. Structure and velocities cause the ray path to follow a non-
straight path down to the horizon and back up, but right at the reflecting surface the
energy path is normal to it.

The reflection point is not directly under the shot point but offset from it; and
after applying normal move out corrections, the source and the receiver are in the
same location.

This section is called "zero offset section," and the ray path is perpendicular to
the dipping reflector. At this point, the time during which the ray traveled to and
from a dipping reflector appears on the section as though the path had been straight
down as in Fig. 5.51(a). Figure 5.51(b) shows the normal incidence principle applied
to the record section to convert to earth model. Note that events moved up-dip after
migration.

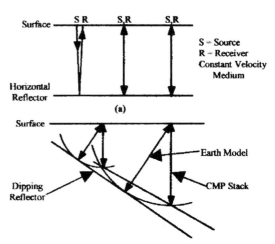

Fig. 5.50 Normal incidence
ray paths

Fig. 5.51 Migration of a
dipping reflection

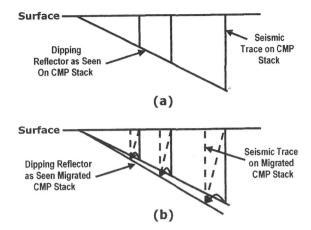

Using the normal incidence principle some subsurface features can be discussed
with regard to how they change when converted from the record section to the earth
model. Some rules can also be formulated on how the features on the record section
change when migrated back to the correct configuration. For simplicity, it shall be
assumed that the acoustic velocity is constant throughout the geologic section, and
that the sections are shot along the direction of dip so that there are no reflections
from one side or the other of the line.

Although the travel paths are normal to the dipping reflectors, the traces are displayed on the record section as if the travel paths were vertical. The purpose of
migration is to move them back up-dip (see Fig. 5.52b). Note that the lateral extent
of the dipping reflector is shortened and the dip angle is steeper after migration.

In Fig. 5.52(a) the anticline is defined by seismic traces displayed vertically and
parallel to each other and the feature appears spread out. Figure 5.51(b) shows the
effect of the migration by applying the normal incidence principle. Events on the
flanks of the anticline moved up-dip cause the lateral extent of the feature to become
smaller. Since the crest of the anticline is horizontal, migration has no effect on it.
The closure (i.e., the maximum height from the crest to the lowest closed contour)
will be the same or a little less.

In Fig. 5.53(a) the syncline is displayed on the record section with seismic traces
vertical and parallel to each other. Figure 5.53(b) shows that the feature becomes

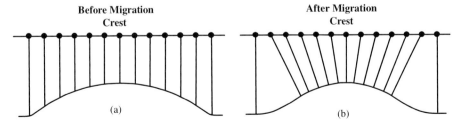

Fig. 5.52 Migration of anticlines. (**a**) record section. (**b**) earth model

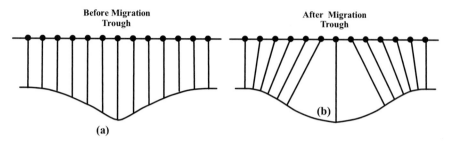

Fig. 5.53 Migration of synclines. (**a**) record section. (**b**) earth model

broader as the ray paths have to reach out to be reflected normally from the reflector after migration. The trough did not move after migration because it is horizontal and the closure is the same or larger.

The amount of relief of the syncline is not very critical, as this structure never has any hydrocarbon accumulation.

Synclines can behave in another way on the record section, if they are relatively narrow or are deep in the section. Deeper or narrower synclines have ray paths that cross on the way down, with one trace being in a position to receive information from two or even three parts of the syncline (Fermat's principal).

The case of two crossing lineups of energy, with perhaps an apparent anticline visible beneath them, is illustrated in Fig. 5.54(b). This is called the *bow-tie effect, or buried focus.* Accordingly, a sharp syncline may be revealed by migration of crossing reflections as shown in Fig. 5.54(a).

When a fault breaks off reflections sharply, or if for some other reason there is a point (or edge) in the subsurface, energy returns from that point to any source within the range. That is, it behaves as a new source. Energy is returned to a number of receivers at different distances from the point source as shown in Fig. 5.55(b). In cross-section, with reflected energy vertically below the shot point, it is an apparent anticline as in Fig. 5.55(a). Sometimes half of it is visible, so the broken-off formation appears to continue in a smooth curve downward.

Although it is not a normal reflection, a diffraction pattern is created by the seismic traces displayed vertically in the record section, so the same process of migration applies. A diffraction pattern is collapsed to a point after migration.

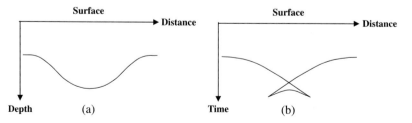

Fig. 5.54 Migration of buried focus (**a**) after migration. (**b**) record section

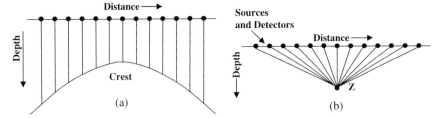

Fig. 5.55 Point source. (**a**) record section. (**b**) point reflector at point Z

Migration Methods

The objective of seismic data processing is to produce an accurate as possible image of the subsurface target, within the constraints imposed by time and money provided. In a few cases the CMP stack, in time or depth, may suffice. In almost every case, today, some sort of migration is required to produce a satisfactory image. There are two general approaches to migration: post-stack and pre-stack. Post-stack migration is acceptable when the stacked data * zero-offset. If there are conflicting dips with varying velocities or a large lateral velocity gradient, *a pre stack partial migration* is used to resolve these conflicting dips.

Pre-Stack Partial Migration (PSPM)

This process, also called *dip moveout* or *DMO*, applied before stack provides a better stack section and an improved migration after stack. Figure 5.56 shows how this occurs. After NMO, the trace is effectively moved to the midpoint position but if there is significant dip the reflection from the dipping reflector is at neither the right place nor the right time. Pre-stack partial migration moves the reflection to the zero-offset point (ZOP). The reflection is still not quite at the right place and time but the

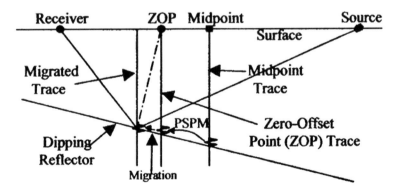

Fig. 5.56 Relationship between zero-offset point and midpoint for a dipping reflector

Fig. 5.57 The conflicting dip
problem

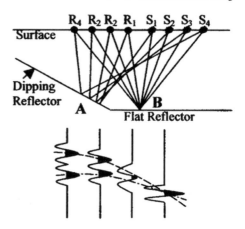

zero-offset assumption of post-stack migration is satisfied. Thus, post-stack migration completes the imaging to the right place and time.

Pre-stack partial migration solves another problem – conflicting dips. See Fig. 5.57.

Two reflections are recorded at about the same time, one from the flat reflector (B) and one from the dipping reflector (A). Reflection A appears to have a higher stacking velocity than B since there is less moveout on A. This is because dip has the effect of making velocities appear to be higher than they actually are. PSPM solves the conflicting dips with different stacking velocities by separating them, since the zero-offset points are different.

The application of pre-stack partial migration can be summarized as follows:

- Data are moved to their zero-offset points, satisfying the condition for post stack migration.
- PSPM solves the conflicting dips with different stacking velocities problem.

Post-Stack Migration Algorithms

There are only two basic approaches used in poststack migration. These are *hyperbolic summation* and *downward continuation.* The Kirchhoff algorithm uses the first approach. The Finite Difference, Stolt, and Gazdag's Phase Shift algorithms all use the downward continuation approach.

Figure 5.58 shows the setting for the harbor example used by Claerbout (1985) to illustrate the physical principle of migration. A storm barrier exists at a distance Z3 from the beach, and there is a gap in the barrier. The gap in the barrier acts as a Huygens secondary source, causing circular wavefronts that approach the beach line (see Fig. 5.59). The gap in the barrier is referred to as *a point aperture.* It is similar to a point source in the subsurface, since both generate circular wavefronts.

In this experiment, it is seen that a Huygens secondary source response to a plane incident wave is to generate an hyperbolic diffraction in the (X,T) plane. See Fig. 5.59.

Fig. 5.58 The harbor setting

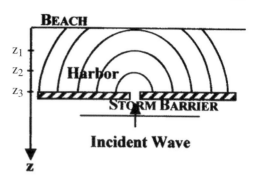

Fig. 5.59 Recording
wavefronts produced by the
storm barrier gap

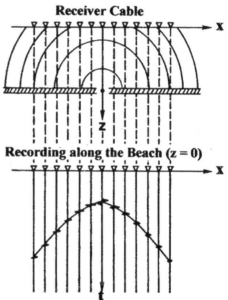

Figure 5.60(a) represents a Huygens source in a depth section. The upper section maps into a point on the zero-offset time section. The vertical axis in the lower section is two-way time. The top of Fig. 5.61 shows multiple Huygens secondary sources and the bottom is the pattern they produce. If the subsurface is assumed to consist of similar source points along each reflecting horizon, the combined wave fronts are similar to reflection wave fronts produced by surface sources. This model behaves much as the gap in the storm barrier. Each of these points acts as a Huygens secondary source and produces hyperbolas in the (X,T) plane. As the sources get closer to each other, superposition of the hyperbolas produces the response of the actual reflecting interface, as shown at the bottom of Fig. 5.62. These hyperbolas are comparable to the diffractions seen at fault boundaries on the stacked sections.

In summary:

Reflectors in the subsurface can be visualized as being composed of many points that act as Huygens' Secondary Sources.

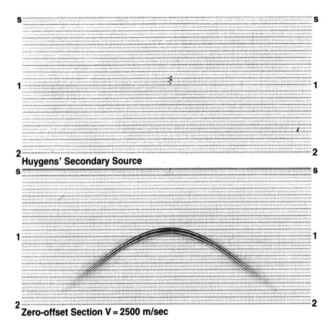

Fig. 5.60 A Huygens secondary source (*top*) and the diffraction it produces (*bottom*)

- The stack section (zero offset) consists of the superposition of many hyperbolic travel time responses
- Discontinuities (faults) along the reflector, diffraction hyperbolas stand out because of diffraction they produce
- A Huygen's secondary – source signature is semicircular in the (X, Z) plane and hyperbolic in the (X, T) plane.

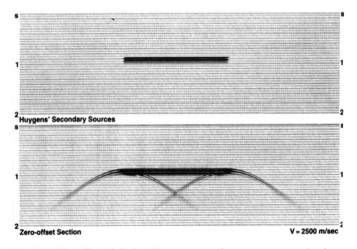

Fig. 5.61 The effect of placing Huygens secondary sources more closely together

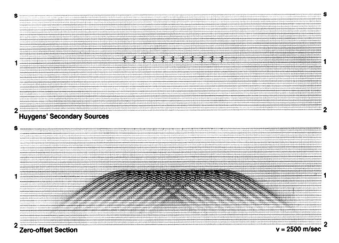

Fig. 5.62 Multiple Huygens secondary sources (*top*) and the diffraction pattern they produce

Figure 5.63 shows the principles of migration based on diffraction summation. Figure 5.63(a) is a zero offset seismic section. (The trace interval is 25 m and constant velocity of 2,500 m/s). Figure 5.63(b) is the migrated version; the amplitude at point B on the hyperbola is mapped onto the apex A along the hyperbolic travel time equation.

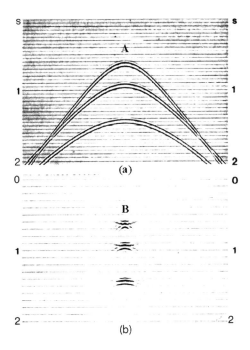

Fig. 5.63 Migration example.
(**a**) zero-offset section.
(**b**) migration

Kirchhoff Migration

Diffraction migration or *Kirchhoff migration* is a statistical approach technique. It is based on the observation that the zero-offset section consists of a single diffraction hyperbola that migrates to a single point Migration involves summation of amplitudes along a hyperbolic path. The advantage of this method is its good performance in case of steep-dip structures. The method performs poorly when the signal-to-noise ratio is low.

Downward Continuation Migration

The harbor model of Fig. 5.58 can also be used to illustrate this approach. The sketch on the far left of Fig. 5.64 shows the diffraction pattern recorded with detectors placed on the beach. The sketches to the left show what would be recorded if the detectors were moved to distances Z_1, Z_2, and Z_3 from the beach. The diffraction gets smaller as the detectors approach the barrier and at Z_3, which is the barrier location, the diffraction collapses to a point. It has been migrated!

Fig. 5.64 Downward continuation based on harbor model

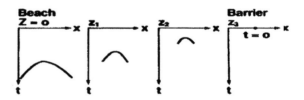

This is the principle of downward continuation migration. Small increments of the stack section are stripped away so that a new time zero, or depth zero, is produced over and over to the bottom of the section. The strips that are removed are appended to the previously removed strips to produce the migrated section. These strips are called time or depth steps. After each downward continuation step, the remaining section must be recalculated. The smaller the step, the more accurately can the section be recalculated.

Finite Difference Migration

This is a deterministic approach that recalculates the section using an approximation of the wave equation suitable for use with computers. One advantage of the finite difference method is its ability to perform well under low signal-to-noise ratio condition. Its disadvantages include long computing time and difficulties in handling steep dips.

Frequency Domain or F-K Domain Migration

Stolt and Phase-Shift migration operate in the F-K domain. Phase shift migration is considered to be the most accurate method of migration but is also the most expensive. It is a deterministic approach via the wave equation instead of using the finite difference approximation. The 2-D Fourier transform is the main technique used in this method. Some of the advantages of *F-K method* are fast computing time, good performance under low signal-to-noise ratio, and excellent handling of steep dips. Disadvantages of this method include difficulties with widely varying velocities.

Migration Examples

Figure 5.65(a) is a stack section. Note that diffractions mask out the true subsurface structure due to the bow-tie effect. Fig. 5.65(b) is a migrated version of the same stack in time domain using the diffraction collapse method, or Kirchhoff migration. Migration unties the bow-ties and turns the structure into a series of synclines.

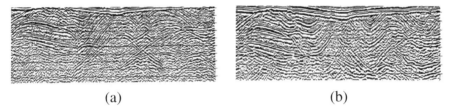

(a)   (b)

Fig. 5.65 "Bow-Tie" example. (**a**) Stack section and (**b**) migration of stack

Figure 5.66 compares a stack section and a time migration section from the Gulf Coast Basin that illustrates a growth fault as seen in (a). Time migration (b) reveals fault planes. Note change in the anticlinal feature better fault information made apparent by the migration. Migration clearly shows the structural complexity on the top of the closed feature by collapsing minor diffractions. Wave-equation time migration was used to migrate the stacked section.

Figure 5.67 is a comparison between a stack section and the F-K migrated version. The migrated section shows more details, diffraction collapse, and excellent fault plane definition.

Time migration is appropriate as long as lateral velocity variations are moderate. When these variations are substantial, depth migration is needed to obtain a true picture of the subsurface. In particular, the geologist prefers to have a depth-migrated section of the subsurface. Unfortunately, due to the lateral change of the velocity and the structure complexity, it is very difficult to get a reliable section.

In the time domain, 3-D migration is needed when the stack section contains events out of the profile. This is the common type of 3-D migration. See Fig. 5.68.

Pre-Stack Migration

Obtaining a satisfactory image of the more complex structures may require pre-stack migration. Both time and depth pre-stack migrations are available, with the latter being the more accurate and expensive. Figure 5.68 illustrates this.

Fig. 5.66 Example of finite
difference time migration.
(**a**) time section.
(**b**) migration of time section

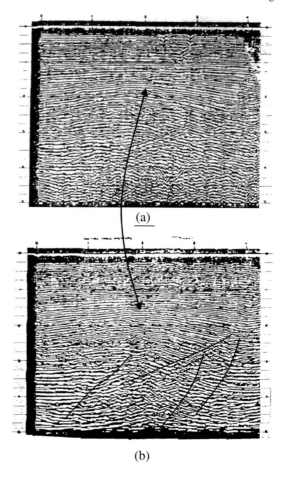

(a)

(b)

Figure 5.69(a) is a model that includes overthrust faults. Synthetic data based on this model were generated and processed two ways. Figure 5.69(b) displays data processed with Kirchhoff DMO, CMP stack, and steep-dip, post-stack migration Fig. 5.69(c) is the same data processed with pre-stack depth migration.

Pre-stack migration requires excellent velocity information. Several iterations may be necessary to obtain optimum results. Note that, if pre-stack migration is done, no intermediate stack is obtained.

Bandpass Filtering

Bandpass filtering is designed to pass signal and reject noise. Filter scans are generated from the data, in which many different narrow, bandpass filters are applied and the results displayed for analysis. The geophysicist designs the final filter to pass the frequencies containing coherent energy (reflections) and to reject those frequencies

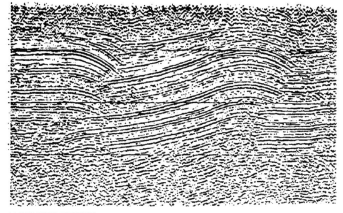

(a) STACKED SECTION

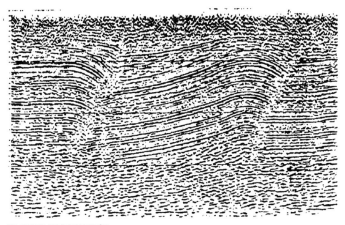

(b) AFTER F-K MIGRATION

Fig. 5.67 Example of F-K migration. (**a**) time section. (**b**) migration of time section

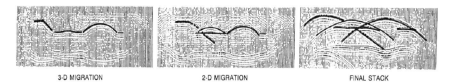

3-D MIGRATION   2-D MIGRATION   FINAL STACK

Fig. 5.68 Migrated and unmigrated 3-D data (Brown, 1991. Reprinted by permission of the American Association of Petroleum Geologists.)

containing mostly noise and no apparent reflections. Time-variant filters are often applied to the data in order to pass higher frequencies shallow in the section and lower frequencies at greater depths. The lines shown on the display of Fig. 5.70 indicate variation of passbands with record time.

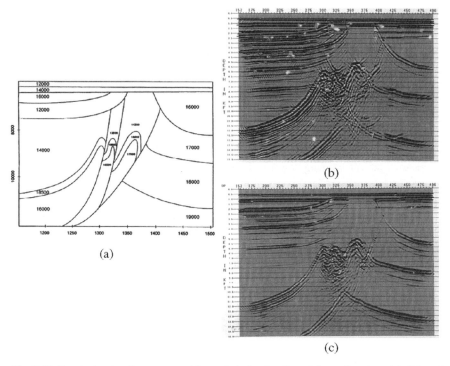

Fig. 5.69 Pre-stack migration compared to post-stack migration. (**a**) overthrust model. (**b**) post-stack migration of model, and (**c**) pre-stack migration of model

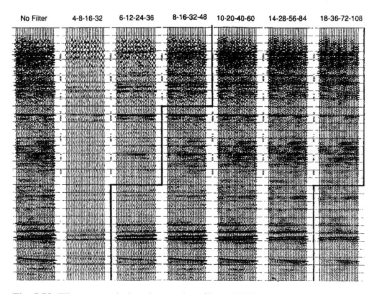

Fig. 5.70 Filter test to design time-variant filters (TVF)

Time-to-Depth Conversion

Record sections are conventionally plotted versus time, as shown on the left of Fig. 5.71. Average velocity functions are used to convert the time axis to a depth axis. On a depth section, usually made after a migration, the wavelengths of the plotted seismic events lengthen with increasing depth. Compare the left and right sides of Fig. 5.71.

This effect is caused in part by the increase of average velocity with depth. Thus, at greater times, equal time intervals correspond to greater depth intervals. The earth's natural filtering action, which results in lower-frequency signals being observed from greater depths even on time sections, accentuates the wavelength stretch. The presentation of seismic signals on a depth scale is more helpful after migration, for diffractions are collapsed and crossing-time dips are separated and moved back to a more nearly correct subsurface spatial position, thus simplifying the structural image. A depth section demonstrates, in geologically meaningful terms, the limitation in the resolution of seismic reflection signals as well as the decrease in the resolution attainable as depth increases. The differences in appearance are quite conspicuous, providing evidence of velocity variation within the section.

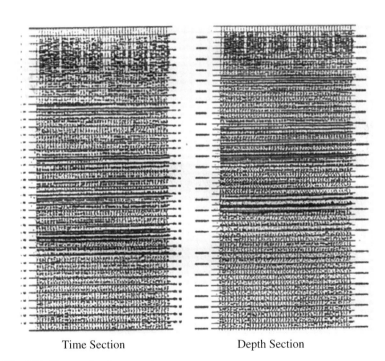

Time Section   Depth Section

Fig. 5.71 Time-to-depth conversion

Displays

The standard black-and-white trace displays have already been introduced. They present amplitude as a function of time and spatial location. Variable-density plots, in which the positive amplitudes are shaded from gray to black according to amplitude, allow a significantly greater range of amplitudes to be discerned than do variable-area plots.

Variable-area and variable-density plots (Fig. 5.72) focus attention primarily on the peaks. A rectified plot (that is, a plot of the absolute values of the trace) allows direct comparison of peaks to troughs. The plot may have both positive and negative values shaded in gray according to amplitude (variable density), or the peaks may be shades of red and the troughs of blue (or any other pair of contrasting colors). A plot in which the peaks are shaded a different color than the troughs, with or without the rectifying of the wiggle trace, is sometimes referred to as a *polarity plot*.

The use of color in the display of seismic data is becoming more important. (See Fig. 5.73.) The color encoding of amplitude onto a wiggle trace (CMP stack) can allow the eye to detect 60 dB or more of amplitude variation, a range far exceeding any black-and-white plot. Substantial increases in the amount of information that can be displayed are possible through the use of color. Actual increases in the dynamic range depend, of course, on the color selection. Grading colors by use of shades of single colors offers the least while sharply contrasting color steps can offer improvements as great as 15-fold. Different color schemes are designed for detecting small amplitude variations over a certain amplitude range or for studying amplitude variation over a broad range of values.

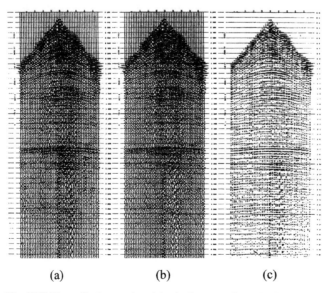

(a)   (b)   (c)

Fig. 5.72 Trace display modes. (**a**) wiggle trace. (**b**) variable area, and (**c**) wiggle trace/variable area (Courtesy of WesternGeco)

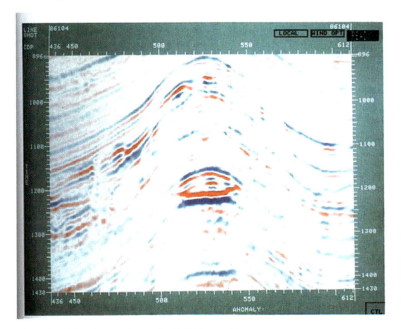

Fig. 5.73 Color display from a Gulf of Mexico Line

Color amplitude plots have the capability to depict a wide range of amplitudes faithfully and with great consistency. Variable-density plots, which are the black-and-white plots capable of conveying the greatest amount of information can vary greatly in quality from day to day and from plotter to plotter due to plotter drift, chemical strength of the developing fluid, and human factors.

The simplest of all conventional plots is a black wiggle-trace plot displaying amplitude as a function of time and space. Color-encoded information displayed on top of a wiggle trace may come from the trace itself (single-channel measurements) or from adjacent channels (multi-channel measurements).

Processing 3-D Data

Almost all concepts of 2-D seismic data processing apply to 3-D data processing, although some complications may arise in 3-D geometry, quality control, statics, velocity analysis, and migration. Editing of very noisy traces, spherical divergence correction for amplitude decay with depth and offset, deconvolution, trace balancing, and elevation statics are done in the preprocessing stage before collecting traces in common cell gathers. Sorting in common cell gathers introduces a problem if there are dipping events, because there are azimuthal variations in the NMO application within the cell.

Data processing must get additional effort on velocity analysis, stacking, and migration. Performed in the preprocessing stage are the steps of editing the bad traces containing high-level noise, geometric spreading correction, deconvolution and trace balancing, and field statics applications (for land data). In conventional 2-D processing, traces are next collected into common-midpoint (CMP) gathers, whereas in 3-D processing traces are collected into common-cell gathers. A common-cell gather coincides with a common-midpoint gather in the case of swath shooting. Sorting into common-cell gathers introduces special problems, such as azimuthal variations of the normal moveout (NMO) within the cell in the case of land geometry, and travel-time deviations from the hyperbolic moveout due to scatter of midpoints within a cell in the case of marine geometry. After NMO and residual statics, CMP stack is performed.

Migration of 3-D data usually involves a first-pass 2-D migration in the in-line direction followed by trace interpolation and a second-pass 2-D migration in the cross-line direction. Some algorithms provide one-pass 3-D migration. One-pass migration most often yields a better migration but run time and cost are greater.

After pre-processing, marine data are ready for common-cell sorting. A grid is superimposed on the survey area. This grid consists of cells with dimensions one-half the receiver group spacing in the in-line direction and nominal line spacing in the cross-line direction. Traces that fall within the cell make up a common cell gather. Due to cable feathering, not all these traces are from the same shot line. The same data processing techniques in 2-D marine apply to 3-D marine data.

Display of 3-D data provides the ability to look at a whole prospect at a particular time or depth via *horizontal slices*. Figure 5.74 is an example of a horizontal time slice.

3-D processing and display can considerably enhance interpretation. Figure 5.75 illustrates this by showing how time slices can be used to draw structure maps.

VSP Data Processing

It is extremely difficult to interpret unprocessed VSP field data. Extensive data processing is required to make VSP recorded data interpretable. The downgoing event is dominant and it is difficult to do any interpretation on the upgoing events. The two event types must be separated. Further, random and coherent noises present in the data must be attenuated.

Separating Upgoing and Downgoing Waves

Figure 5.76 shows raypaths for both primary and multiple reflections. The reflectors in this analysis are assumed to be flat and horizontal. The source is actually located near the well and rays are nearly vertical. The horizontal distance is exaggerated so that the rays are separated for visual clarity. $T_1 t T_2$, and T_G are one-way vertical

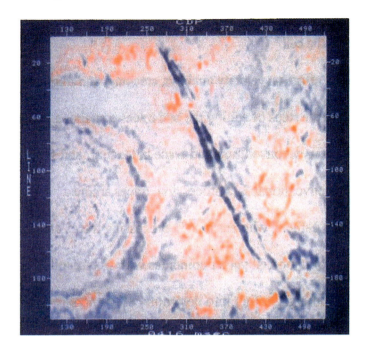

Fig. 5.74 Horizontal section or time slice from the Gulf of Mexico

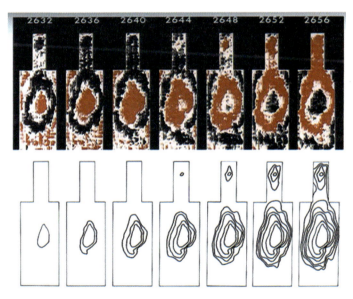

Fig. 5.75 Structure map derived from sequence of time slices 4 ms apart (Courtesy of Occidental Exploration and Production Company)

Fig. 5.76 Upgoing primaries and multiples

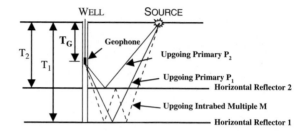

travel times to reflectors 1 and 2, and geophone depth, respectively. If the time for upgoing primary P_1 is t_{P1}, the time for upgoing primary P_2 is t_{P2}, and the time for upgoing multiple M is t_{Mu}, it can be seen from Fig. 5.75 that:

$$t_1 = 2T_1 - T_G,$$
$$t_2 = 2T_2 - T_G, \text{ and}$$
$$t_M = 2T_1 + 2(T_1 - T_2) - T_G,$$

Raypaths for downgoing events and intrabed multiples are illustrated in Fig. 5.77. If the time for downgoing multiple M_1 is t_{M1}, the time for downgoing multiple M_2 is t_{M2}, and the time for downgoing multiple M is t_{Md}, it can be seen from Fig. 7.11 that:

$$t_{M1} = 2T_1 + T_G,$$
$$t_{M2} = 2T_2 + T_G, \text{ and}$$
$$t_M = 2T_1 - 2(T_1 - T_2) - T_G,$$

From the travel time equations, a static time shift of either $+T_G$ or $-T_G$ will position VSP events at the same times at which they would be recorded by a surface geophone at position G. By definition, T_G is the first-break time for the VSP trace recorded at geophone position G, so if T_G is added to both sides of the upgoing traveltime equation and subtracted from both sides of the downgoing equation, the upgoing and downgoing events can be separated. See Fig. 5.78.

While VSP data contain both downgoing and upgoing waves, analysis of upgoing waves is particularly important because these events correspond to the ones recorded by surface seismic surveys. Thus, processes that attenuate the downgoing modes without seriously affecting upgoing events are very important in VSP data processing. The most common technique used to remove a selected VSP wave mode is F-K or velocity filtering.

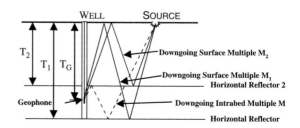

Fig. 5.77 Downgoing surface and intrabed multiples

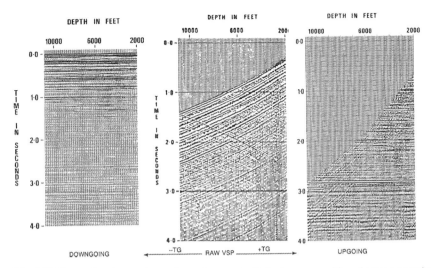

Fig. 5.78 Separating downgoing and upgoing events. Reflectors are positioned at two-way times

Since this process is applied in frequency-wavenumber (F-K) space, it is important to avoid temporal or spatial aliasing. This requires that VSP data be recorded at uniform increments in both time and space that satisfy the Nyquist sampling theorem. Figure 5.79 illustrates the F-K velocity filtering applied to enhance the desired wave modes. Note that the downgoing P-wave (P) and S-wave (S) events are in the positive side of the F-K plane while the upgoing P-wave (P') and S-wave (S') events are in the negative side of the F-K plane. Length of the lines in Fig. 5.79 is proportional to the energy contained in the wave mode.

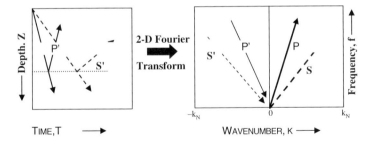

Fig. 5.79 Velocity filtering

Summary and Discussion

There is a belief that the powerful computers and the sophisticated software applications can solve field problems. The bad news is that the computer software will not solve all the problems. If the problem of the data is resolution and insufficient data acquired in the field, or errors in the field parameters or malfunctions of the field equipment, computer software cannot retrieve such information because it is not present in the field seismic data.

Data processing is a very powerful tool in manipulating the field information. It converts the raw data with all information, including noise and distorted information, to meaningful seismic cross-sections that represent vertical slices through the subsurface. These slices represent geological information and potential hydrocarbon traps.

A generalized seismic data-processing flow has been presented. The intention is to explain the physical meaning of some "buzz" words that are heard, such as *Cross-correlation* that is used to extract information from seismic data acquired by the Vibroseis method; it is used to solve statics problems and for many more seismic data-processing applications. *Gain* is applied to compensate for spherical divergence and to preserve the relative true amplitude, which can be used as hydrocarbon indicator.

Convolution is used to filter out some undesirable frequencies while *deconvolution* is a process to attenuate short-period multiple reflections and to enhance the vertical resolution. *Normal moveout corrections* are applied to correct for the field geometry and produce a zero-offset stack section as if the source and receiver were in the same location.

Elevation statics are applied to minimize the effect of the variation of the surface elevations. *Refraction statics* are applied to solve the irregularities in the near-surface formation called. "weathering layer" or "near-surface layers." It may vary in thickness and velocities, both laterally and vertically. These near-surface layers distort the deep structural and stratigraphic features that are obtained from seismic sections.

Data processing is both an art and science. The processing steps should be handled in a logical order. Every step should be checked carefully before proceeding to the next step. Remember that the goal of processing the seismic data is to obtain geologically sound sections and not just a pretty one with man-made information.

The map is only as reliable as the seismic data. Many dry holes have been drilled on what appeared to be structural highs due to processing problems. As was discussed in Chap. 4, it is difficult to calculate the cost of acquiring a mile of seismic data. It also is not easy to give a price range for processing a mile of seismic data. The cost per mile depends upon the type of data, whether it is marine or land, the signal-to-noise ratio of each type, the field problems involved, the field configuration such as the number of channels, fold, and sampling interval; and the processing sequence needed including special software programs to perform certain tasks.

If a contractor is used to process newly acquired seismic data or to evaluate a prospect and a reprocessing job is needed, it is advisable to consult with several data processing contractors. Software applications for basic processing are very similar from one contractor to another. The knowledge and the experience of the processing geophysicist is the major factor in the quality and the validity of the processed data.

The interaction of all staff members in the data processing sequence is very important. Seismic workstations may be used for post-stack enhancement programs such as modeling, synthetic seismograms, 3-D interpretation, and other applications. The cost of having one of these stations varies according to the hardware configuration required to apply the various software packages. It varies from $10,000 for a small system to $200,000 for a powerful hardware configuration.

Migration is a process that moves seismic events to their proper subsurface positions due to the way record sections are displayed. After applying normal moveout corrections, it is assumed that the seismic section is a series of arrival times with the normal incidence ray paths at the reflectors. This section is called zero-offset because the NMO process moves the source and the receiver at the same position.

In the case of flat, or a nearly flat horizons, the record section (seismic data display) and the earth model are the same. In case of dipping reflectors, they are not the same. The normals to the reflectors are displayed vertically and parallel to each other; their true subsurface positions are rotated down from their original position.

Migration is performed to move these reflectors to their proper positions so that the record section matches the earth model. The process of migration moves events up dip and makes the ray travel path normal to the reflectors. It shortens the lateral extent of the event, and the angle of dip will be steeper after migration. By applying this principle, one can observe that an anticline will have a narrower areal extent after migration while a syncline will have a broader areal extent. The crest of the anticline and the trough of the syncline do not move because their position is horizontal, while the closure in both cases may not change.

Diffractions from the top of the fault plane are observed as hyperbolas. They collapse to a point after migration. This will help the interpreter to more accurately delineate and orient fault patterns. In case of a deep sharp syncline (buried focus), migration unties the bow-tie feature on the record section and reveals the true synclinal features.

Migration improves the horizontal resolution and gives more accurate subsurface picture of the geologic structures, which give more realistic size of the hydrocarbon accumulations in place.

There are different methods of migration. Each of which is designed to handle a different geologic setting of the target structure.

Exercises

1. The maximum frequencies for the data sets A, B, and C are:

    A. 75 Hz
    B. 61 Hz
    C. 129 Hz

    What sample intervals can be used for each data set?

2. Find the cross-correlation of the following wavelets: $A.(3, -4, 2, 1)$ $B.(1, 0, -6, 2)$.
3. Find the convolution of $A.(4, -2, 1, 3)$ and $B.(-1, 0, 1)$.
4. Why must gain be applied to seismic records?
5. What is the difference between a front-end mute and a surgical mute? Why are they applied?
6. Define the following terms: a. Stacking velocity, b. Average velocity, and c. Migration velocity.
7. From the listed offsets and reflection times given below, do an $X^2 - T^2$ plot. Fit a straight line to the plotted points and estimate velocity, V, and zero-offset time, T_0.

Offset, X (m)	Time, T (s)
100	2.251
500	2.255
900	2.265
1300	2.287
1700	2.300
2100	2.338
2500	2.380
2900	2.410
3300	2.460
3700	2.500
4100	2.575
4500	2.620
4900	2.700
5300	2.760
5700	2.830

8. What is the purpose of a velocity filter?
9. How does whitening or spiking decon differ from gapped decon?
10. Why design and apply more than one decon filter to seismic traces?
11. Do bandpass filters get rid of all the noise on a seismic record or section? Why or why not?
12. Why do both refraction-based residual statics and reflection-based residual statics?
13. Why run multiple iterations of velocity and residual statics analysis?
14. CMP stack combines all traces that have common midpoints into single trace. Can these midpoint traces always be considered to be zero-offset traces, as well? If not, what process can be used to produce zero-offset traces?
15. Does post-stack migration cause reflection dip to increase or decrease compared to the stack?
16. In what way does time-to-depth conversion improve imaging of the geologic target?

Bibliography

Angeleri, G. P. and E. Loinger. "Amplitude and Phase Distortions Due to Absorption in Seismograms and VSP." Paper presented at 44th annual meeting of EAEG, (1982).

Baysal, E., D. D. Kosloff and J. W. C. Sherwood. "Reverse Time Migration." *Geophysics, 48* (1983):1514–1524.

Berkhout, A. J. *Seismic Migration—Imaging of Acoustic Energy by Wave Field Extrapolation.* Amsterdam, Netherlands: Elsevier Science Publ. Co., Inc., (1980).

Black, J. L., I. T. McMahon, H. Meinardus and I. Henderson. "Applications of Prestack Migration and Dip Moveout" *Paper presented at the 55th Ann Int. Soc. Explor. Geophys. Mfg.,* (1985).

Brown, A. R. "Interpretation of Three-Dimensional Seismic Data." *Memoir 42* (1988).

Brown, A. R., "Interpretation of Three-Dimensional Seismic Data", Third Edition, Tulsa. Oklahoma, U.S.A., AAPG, (1991).

Chun, J, H. and C. Jacewitz, "Fundamentals of Frequency-Domain Migration", *Geophysics 46* (1981):717–732.

Claerbout, J. F. *Fundamentals of Frequency-Domain Migration.* New York: McGraw-Hill, (1976).

Claerbout, J. F. *Imaging the Earth's Interior.* Oxford: Blackwell Scientific Publications, (1985).

Claerbout, J. F. and S. M. Doherty. "Downward Continuation of Moveout-Corrected Seismograms." *Geophysics' 37* (1972):741–768.

Dalley, R. M. et al. "Dip and Azimuth Displays for 3-D Seismic Interpretation." *First Break.* (March 1989):86–95.

Dunkin, J. W. and F. K. Levin. "Isochrons for a Three-Dimensional System." *Geophysics 36* (1971):1099–1137.

Enachescu, M. "Amplitude Interpretation of 3-D Reflection Data." *TLE12* (1993):678–685.

Greaves, R J. and T. J. Flup. "Three-Dimensional Seismic Monitoring of an Enhanced Oil Recovery Process." *Geophysics 52* (1987):1175–1187.

Fowler, P. "Velocity-Independent Imaging of Seismic Reflectors." *Presented at the 54th Ann. Int. Soc. Explor. Geophys. Mtg.,* Atlanta, December, (1984).

Gardner, G. H. F., W. S. French, and T. Matzuk. "Elements of Migration and Velocity Analysis." *Geophysics 39* (1974):811–825.

Gadzag, J. "Wave-Equation Migration by Phase Shift." *Geophysics 43* (1978):1342–1351.

Gadzag, J. and P. Squazzero. "Migration of Seismic Data by Phase Shift Plus Interpolation." *Geophysics 49* (1984):124–131.

Hileman, J. A., P. Embree, J. C. Pfleuger,. "Automatic Static Corrections", *Geophysical Prospecting 16* (1968).:328–358.

Hilterman, F. J. "Three-Dimensional Seismic Modeling." *Geophysics 35* (1970):1020–1037.

Hilterman, F. J. "Amplitudes of Seismic Waves. A Quick Look." *Geophysics 40* (1975):745–762.

Hubral, P. and T. Krey. "Interval Velocities from Seismic Reflection Time Measurements." *Society of Exploration Geophysicists Monograph* (1980).

Kluesner, D. F. "Champion Field: Role of Three-Dimensional Seismic in Development of a Complex Giant Oilfield." *AAPG Bulletin 72* (1988):207.

Lee, M. W. and A. H. Balch. "Computer Processing of Vertical Seismic Profile Data." *Geophysics 48* (1983):272–287.

Lee, M. W. and S. H. Suh. "Optimization of One-Way Wave Equations." *Geophysics 50* (1985):1634–1637.

Levin, F. K., "Apparent Velocity from Dipping Interface Reflections." *Geophysics 36* (1971): 510–516.

Lindsey, J. P. "Elimination of Seismic Ghost Reflections by Means of a Linear Filter." *Geophysics' 25* (1960):130–140.

Mayne, W. H. "Common Reflection Point Horizontal Stacking Techniques." *Geophysics 21* (1962):927–938.

Mufti, I. R., J. A. Pita, and R. W. Huntley, "Finite- Difference Depth Migration of Exploration-Scale 3-D Seismic Data," *Geophysics 61* (1996):776–794.

Neidell, N. S. and M. T. Taner. "Semblance and Other Coherency Measures for Multichannel Data." *Geophysics 36* (1971):482–497.

Nestvold, E. O. "3-D seismic: Is the Promise Fulfilled." *TLE 11* (1992):12–19.

Newman, P. "Divergence Effects in a Layered Earth." *Geophysics 38* (1973):481–488.

Osman M. O. "Discrimination Between Intrinsic and Apparent Attenuation in Layered Media." Master of Science Thesis. The University of Tulsa, Tulsa, OK, (1988).

Palmer, D. "The Generalized Reciprocal Method of Refraction Seismic Interpretation." *Geophysics 46* (1981):1508–1518.

Pullin, N., L. Matthews, and K Hirsche. "Techniques Applied to Obtain Very High Resolution 3-D Seismic Imaging at an Athabasca Tar Sands Thermal Pilot." *TLE 6* (1987): 10-15.

Reauchle, S. K., T. R. Earr, R. D. Tucker, and M. T. Singleton. "3-D Seismic Data for Field Development: Land Side Field Case Study." *TLE 10* (1991):30–35.

Reblin, M. T., G. G. Chapel, S. L. Roche, and C. Keller. "A 3-D Seismic Reflection Survey over the Dollarhide Field, Andrews County, Texas." *TLE 10* (1991):11–16.

Ritchie, W. "Role of the 3D Seismic Technique in Improving Oilfield Economics." *Journal of Petroleum Technology* (July 1986):777–786.

Robertson, J. D. "Reservoir Management Using 3D Seismic Data." *Journal of Petroleum Technology* (1989):663–667.

Robinson, E. A. "Dynamic Predictive Deconvolution." *Geophysical Prospecting 23* (1975):779–797.

Robinson, E. A., and S. Treitel. *Geophysical Signal Analysis.* Englewood, N. J.: Prentice-Hall, (1980).

Robinson, E. A. *Migration of Geophysical Data.* Boston: IHRDC, (1983).

Robinson, E. A. *Seismic Velocity Analysis and the Convolutional Model.* Boston: IHRDC, (1983).

Rothman, D., S. Levin and F. Rocca. "Residual Migration: Applications and Limitations." *Geophysics, 50* (1985):110–126.

Schneider, W. A. "Developments in Seismic Data Processing and Analysis (1968–1970)" *Geophysics 36* (1971):1043–1073.

Schneider, W. "Integral Formulation for Migration in Two and Three Dimensions." *Geophysics 43* (1978):49–76.

Schneider, W. A. and S. Kuo. "Refraction Modeling for Static Corrections." *55th Ann. Int. Soc. Explor. Geophys. Mtg.* (1985).

Sheriff, R. E. "Encyclopedic Dictionary of Exploration Geophysics." *Society of Exploration Geophysicists* Tulsa, OK, (1973).

Sheriff, R. E. *A First Course in Geophysical Exploration and Interpretation.* Boston: IHRDC, (1978).

Sherwood, J. W. C. and P. H. Poe. "Constant Velocity Stack and Seismic Wavelet Processing." *Geophysics 37* (1972): 769–787.

Stolt, R. H. "Migration by Fourier Transform." *Geophysics 43* (1978):23–48.

Taner, M. T. and F. Koehler. "Velocity Spectra." *Geophysics 34* (1969):859–881.

Taner, M. T., F. Koehler, and K A. Alhilali. "Estimation and Correction of Near-Surface Time Anomalies." *Geophysics 39* (1974):441–463.

Tatham, R. H. and P. Stoffa, "A Potential Hydrocarbon Indicator." *Geophysics 41* (1976):837–849.

Telford, W. M., L. P. Geldart, R. E. Sheriff, and D. A. Keys. *Applied Geophysics.* Cambridge, England: Cambridge University Press, (1976).

Walton, G. G. "Three-Dimensional Seismic Method." *Geophysics 37* (1972):417–430.

Waters, K H. *Reflection Seismology.* New York: John Wiley, (1978).

Wiggens, R. A., K. L. Lamer, and R. D. Wisecup. "Residual Static Analysis as a General Linear Inverse Problem." *Geophysics to.* (1976):992–938.

Yilmaz, O. and J. F. Claerbout. "Prestack Partial Migration." *Geophysics 45* (1980):1753–1777.

Yilmaz, O. and R. Chambers. "Migration Velocity Analysis by Wave Field Extrapolation." *Geophysics 49* (1984):1664–1674.

Yilmaz, O. "Seismic Data Processing." Tulsa: OK: *Society of Exploration Geophysicists,* (1987).

Chapter 6
Seismic Interpretation

Introduction

Seismic interpretation provides an assessment of a prospect's hydrocarbon potential and, if favorable, identifies best locations for drilling wells. Interpretation should make use of all of the following that are available:

- Vertical seismic sections (usually migrated)
- Horizontal seismic sections
- Velocity models
- Well logs
- VSP data
- Amplitude versus offset (AVO) analyses
- Geochemical analyses
- Other information obtained from previous drilling such as the presence **of** high pressure zones in the subsurface

This chapter discusses the following techniques used in seismic interpretation:

- Modeling
- Tomography
- AVO
- VSP interpretation

Modeling

The most frequent use of models in seismic exploration is to check interpretations of seismic data or to do initial interpretation, often via interactive systems. Sometimes experiments are performed on models constructed to scale using materials of known physical properties to duplicate observed data. Mathematical models are more often used. These are designed and stored in computers.

There are two basic types of seismic modeling – forward and inverse. In both types of modeling, parameters such as layer densities, layer thicknesses, interval

M.R. Gadallah, R. Fisher, *Exploration Geophysics*,
DOI 10.1007/978-3-540-85160-8_6, © Springer-Verlag Berlin Heidelberg 2009

velocities, and a number of rock layers, N, are selected The principal difference between the two is that forward modeling parameters are selected independently of seismic data while inverse modeling infers parameters from recorded seismic data. Initial model parameters in forward modeling may be based on geophysical and geological data obtained from borehole measurements or selected randomly from a set of plausible values.

In both methods, the selected parameters are used to generate a synthetic seismic trace, section, record, or even a 3-D data volume. The synthetic data are compared to the recorded seismic data. An error function, usually least square error, is determined. If this error exceeds the desired maximum, model parameters are modified and the calculations are repeated. Frequently, several iterations are required to achieve desired agreement between synthetic and recorded seismic data. Several earth models may be generated and synthetic data generated from each. Successive iterations reduce the number of earth models generated and narrow the range of parameter values until a single model that produces synthetic data within the desired "closeness of fit".

As discussed in the migration section of Chap. 5, seismic data are subject to distortion. The earth models used, or developed, must contain the mechanisms that cause the observed distortions. Consequently, a review of those distortions observed on seismic reflection exploration measurements follows, below.

Figure 6.1 shows how modeling can lead to a more accurate image of the subsurface. The seismic section at the top left is obviously a distorted image of the subsurface. The interpreter sees this distorted view of the subsurface on the section and attempts to compensate for the distortion mechanisms. The initial interpretation (top right of Fig. 6.1) shows a faulted anticline and a potential reservoir below the

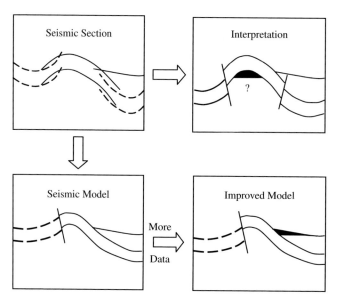

Fig. 6.1 Enhanced imaging through modeling

apes of the anticline. Modeling produces the image at the lower left of Fig. 6.1. This casts doubt on the fault on the right side of the anticlinal feature. Additional seismic and geological data and studies lead to the improved model at the lower right of Fig. 6.1. A well drilled on the initial interpretation would probably have been dry. Modeling improved the probability of a producing discovery well.

Seismic Distortions

Three types of seismic distortions are of most concern:

1. Events are not shown in their true positions, in time or depth,
2. Reflection response differences cause apparent bedding changes.
3. Noise (extraneous events) interferes with and masks signal.

Mechanisms that produce distortion include:

- Focusing
- Shadow zones
- Subsurface discontinuities
- Inadequate resolution
- Topography and near-surface variations
- Velocity changes
- Noise generators

Focusing

Figure 6.2 illustrates distortion of irregular and dipping beds on CMP stack sections. Both an anticline and a syncline are shown. The reflected energy fanning out from the convex surface of the anticline makes it appear larger than its true dimensions. By contrast, the syncline is made narrower than its true size.

In Figure 6.3, the zero offset section shows the bow tie effect caused by the two deep-seated synclines. The process of migration removes the distortion in the stacked section, showing three synclines and one anticline. The bow ties were untied, revealing the true subsurface structure as synclines. The anticline is reduced in size.

The process of migration removes the distortion in the stacked section but optimum migration velocities are vital to obtaining an optimum migration. Other factors may also limit the effect of migration.

Shadow Zones

The section of Fig. 6.4 has areas with no reflections (dead areas). These are called shadow zones and are common in the vicinity of faults and other discontinuous areas in the subsurface. Shadow zones result when energy from a reflector is focused on receivers that produce other traces. As a result, reflectors are not shown in their true positions.

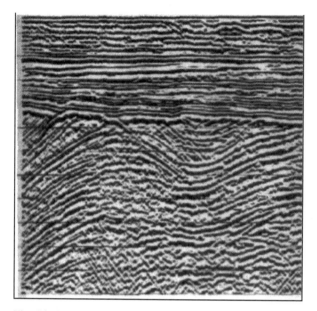

Fig. 6.2 Focusing in anticlines and synclines. Courtesy WesternGeco

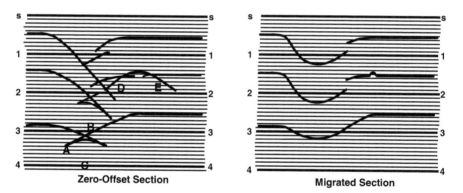

Fig. 6.3 Bow tie effect of buried focus

Subsurface Discontinuities

Diffractions occur at discontinuities in the subsurface such as faults and velocity discontinuities (as at "bright spot" terminations). Figure 6.5 shows a CMP stack section for a seismic line across a large horst block. Diffractions mask the fault plane on the left of the section. The normal incidence ray path model of Fig. 6.6 shows this feature without diffractions. The fault zone can be seen more clearly. A shadow zone in the fault vicinity can also be seen.

Fig. 6.4 Shadow zones

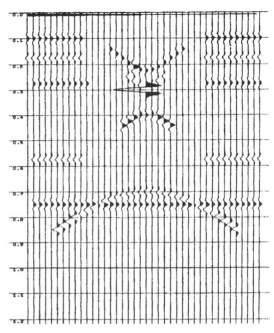

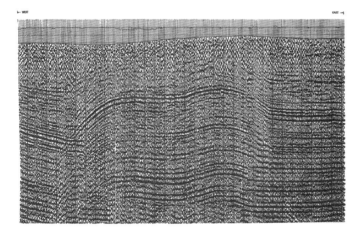

Fig. 6.5 CMP stack section for a line across a block formed by reverse faulting

Inadequate Resolution

Only the lower end of the acoustic spectrum is available to the seismic method because the earth attenuates higher frequencies much more than lower frequencies. In most cases, the usable frequency range does not exceed 80 Hz. Often lower limits are imposed by ground roll and wind noise that include the lower and higher

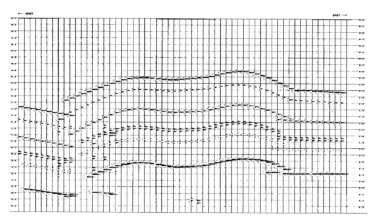

Fig. 6.6 Normal incidence ray path model of fault block shown in Fig. 6.5

frequencies of the 5–80 Hz range. The "signal" frequency band is more often on the order of 20–60 Hz.

Bandwidth limits produce limits on resolution. When layers become thin with respect to wavelength (about 100 ft), the wavelets from the top and bottom of the layer interfere with one another. This results in the appearance of a single complex wavelet rather than two separate wavelets. For very thin layering, the interference becomes so severe that only one strong reflection or no reflection is present as in Fig. 6.7.

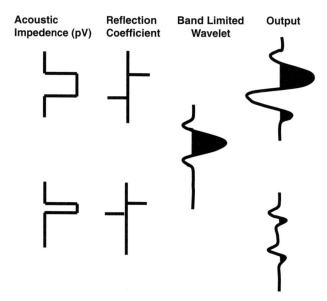

Fig. 6.7 Thin bed response

Topography or Near Surface Variations

Variations in the surface and near surface frequently cause seismic distortion. Changes in surface material cause source and receiver response variations that affect reflection quality. Near surface velocity and thickness variations, laterally and vertically, can cause apparent structure in the subsurface as in Fig. 6.8. Here the model is a flat horizon overlain by a constant thickness near surface with extreme lateral velocity variation. Travel paths through the higher velocity result in earlier reflection times while paths through the lower velocity result in later reflection times. As a result the reflection from a flat horizon appears to have structural variations.

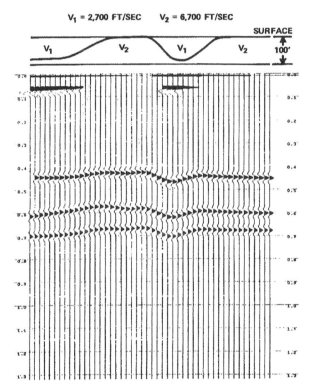

Fig. 6.8 Distortion in the seismic data because of lateral near surface velocity variation

Removing time shifts such as those in Fig. 6.8 before stacking data is the objective of static correction programs. Various methods are used for this but most are unable to eliminate time variations caused by gradual variations in a near surface layer, requiring special or unusual approaches in some complex areas.

Velocity Changes

Change in velocity across boundaries in the subsurface produces reflections but some kinds of velocity variation can distort the view of the subsurface. Examples include:

1. Velocity pull-up from shallower beds
2. Dipping bed effects
3. Overlying anomalies
4. Over-pressured shale zones

Figure 6.9 is an example of velocity pull-up. A shallow salt body produces anomalously low reflection times where raypaths pass through it because its velocity is higher than the sand and shale that surround it. This can be readily identified if there is a sequence of flat reflectors below it, but not so easily if the time anomalies are superposed on real structure.

Figure 6.10 is a CMP stack section on which a series of dipping reflectors appear to thin from left to right. The model of Fig. 6.11 shows the true situation. The dipping beds are of uniform thickness but the increasing thickness of the overlying sediments cause the stacking velocity to change laterally, producing apparent thickness changes as well as false dips.

Figure 6.12 is an earth model with an abrupt lateral velocity change from 10,000 fps to 16,000 fps. Figure 6.13 traces rays from the two velocity discontinuity endpoints. Figure 6.13 shows a zero-offset section produced from the model of Fig. 6.12 that shows a false or pseudo fault.

Figure 6.14 is an earth model with over-pressured shale. The result is a decrease in velocity in the over-pressured zone. Since the shale's upper surface is anticlinal the distortion from the top of the shale is not surprising. However, the reflection from the bottom of the shale is also distorted because of the lateral velocity change. See Fig. 6.15.

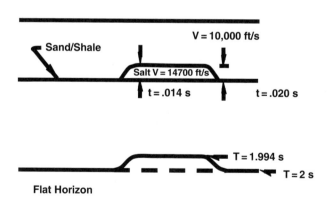

Fig. 6.9 Velocity pull-up

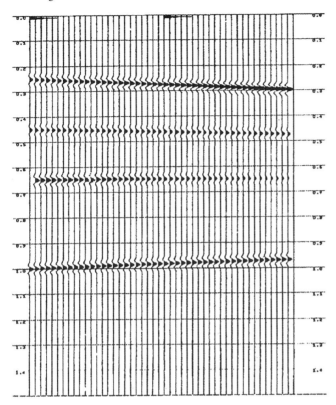

Fig. 6.10 Subsurface section – basin-ward thinning

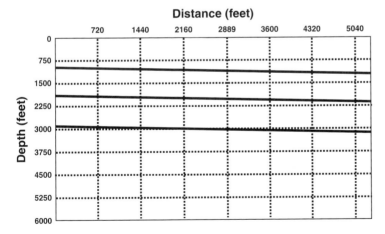

Fig. 6.11 Seismic model – basinward thinning

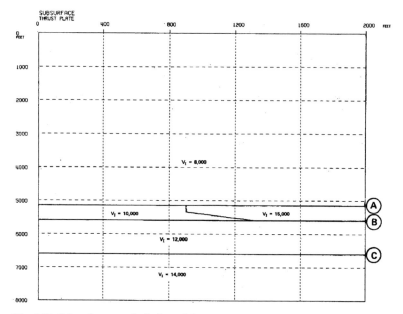

Fig. 6.12 Subsurface pseudo fault model

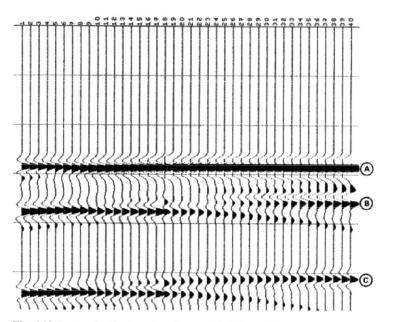

Fig. 6.13 Ray tracing for the subsurface model of Fig. 6.12

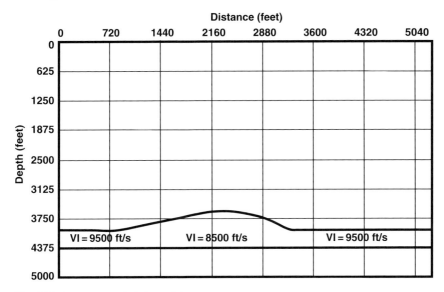

Fig. 6.14 Over-pressured shale model

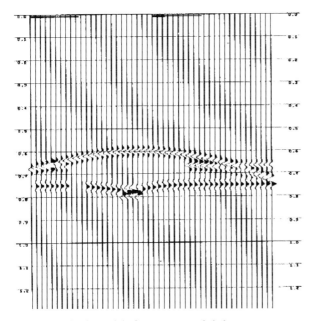

Fig. 6.15 Seismic model of over-pressured shale

Noise

Most advanced software and hardware in use today provide optimum acquisition and processing of seismic data but residual noise will mask the subsurface to some extent. Examples of noise are:

- Random Noise
- Residual Coherent Noise
- Out of Plane Reflections
- Multiples
- Ghosts

This interfering energy can cause apparent reflection changes in time (pseudo faults), frequency (pseudo stratification change), and amplitude (apparent change in reflection coefficient), or phase (structure or stratigraphic anomaly).

The distortion mechanisms, discussed above, as well as others present in seismic data must be recognized, and effects removed to the extent possible. The modeling process is another way of doing this.

One Dimensional Modeling

Synthetic seismic traces are one-dimensional models. Data for generating them come from interval transit time logs or continuous velocity logs (Fig. 6.16) and density logs.

The method of calculating synthetics is as follows:

1. Integrate a velocity log (velocity versus depth series) to produce a velocity versus time log.
2. Convert a density log of the same well to density versus time.
3. Combine the data of the above two logs to obtain a reflection coefficient versus time log,
4. Include, as desired:

   a. Transmission losses.
   b. Multiples (first order or all multiples)
   c. Ghosts
   d. Reverberations.
   e. Noise

5. Convolve the reflection coefficient log with a wavelet of choice to obtain the synthetic seismogram.

The result is a model based on a well log that closely resembles a seismic trace recorded at the same location.

Comparison with the synthetic allows correlation of events on seismic sections to geologic markers. These markers are easily seen on the log. Figure 6.17 shows a

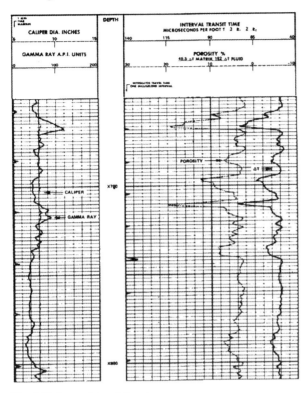

Fig. 6.16 Interval transit time log

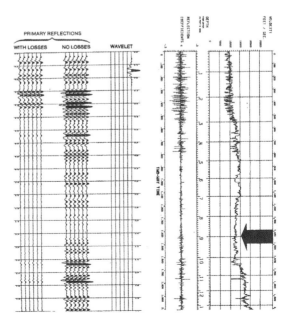

Fig. 6.17 Primary reflection synthetic

synthetic seismogram (a set of synthetic traces). Synthetics can also be used in selecting acquisition and processing procedures that best preserve data most important
to the prospect.

Modeling can be used to test the validity of assumptions regarding the effect of
changes in the geologic section. Geologic changes that can be modeled include:

1. Bed thickness
2. Bed velocity
3. Bed density
4. Slabs or ramps of velocity or density in logs

Merging logs is another way geologic changes can be effected.

Figures 6.18 , 6.19 and 6.20 demonstrate, with logs, the change in seismic response as geology changes. In each case an arrow indicates the part of the log
that is modified. Another application of synthetic seismograms is demonstrated by
Figs. 6.21 and 6.22. Five wells were selected for the study. Interval velocities plotted
against time for each of the five wells provide the model, as shown in Fig. 6.21. The
flat event D, representing an ancient seabed is obtained by time shifting the logs.
High-speed sandstone formation B-C represents a deltaic deposition with most of
the material near shore to the right, as it would have been deposited under the simulated condition. Figure 6.22 shows synthetic seismogram traces for the five wells.
Amplitude and phase distortion in the pinchout zone of B-C is quite apparent.

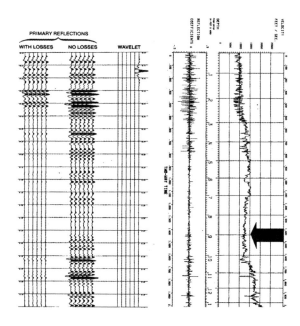

Fig. 6.18 Primary reflection
synthetic with velocity
modified between 8700 and
9350 ft

Two Dimensional Modeling

One-dimensional modeling techniques can be used to study the subsurface below discrete points on the surface. Two-dimensional modeling is required where

Fig. 6.19 Primary reflections
synthetic with depth
modification At 8700 ft. Bed
thickness was reduced from
430 to 312 ft

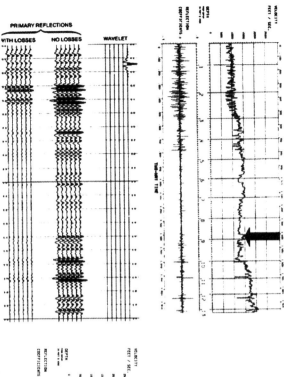

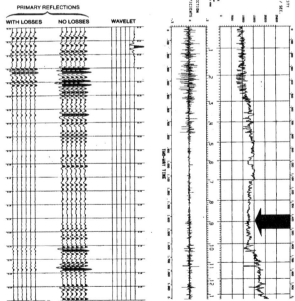

Fig. 6.20 Primary reflections synthetic with repeat section to simulate thrust faulting. The section
begins at 8700 ft (Fig. 6.20) was edited into this log at depth beginning at 7850 ft

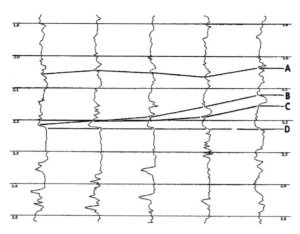

Fig. 6.21 Model cross section – interval velocity versus time

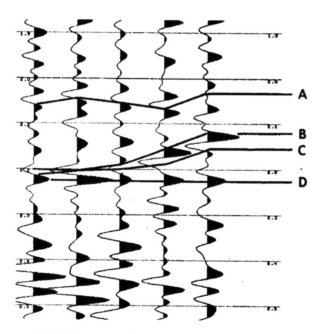

Fig. 6.22 Model cross section-primary reflection

there are lateral variations between points that cause focusing problems, shadow zones, diffractions, etc. There is a wide range of two-dimensional models. Two-dimensional models are used to confirm or reject geologic interpretations of seismic data. Complexity of the modeling approach, and cost of the modeling, varies with complexity of the subsurface. Normal incidence ray tracing is the most widely used modeling technique because it gives the desired accuracy for most problems at a reasonable cost.

Steps in ray-tracing modeling are listed below:

- Assume or derive parameters that describe the assumed subsurface geology. This should include bedding geometry, interval velocities, and densities (if available). See Fig. 6.23.
- Input the model parameters into a computer in the form required by the software being used
- Perform ray-tracing calculations (Fig. 6.24)
- Determine vertical incidence reflection and transmission coefficients for every interface in the model
- Generate a unit impulse seismic section (Fig. 6.25a)
- Convolve the selected source wavelet with the unit impulse section (Fig. 6.25b)
- Add random noise if this is appropriate to the problem (Fig. 6.26)

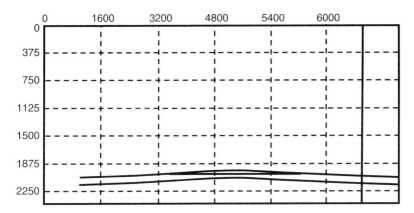

Fig. 6.23 Subsurface depth model

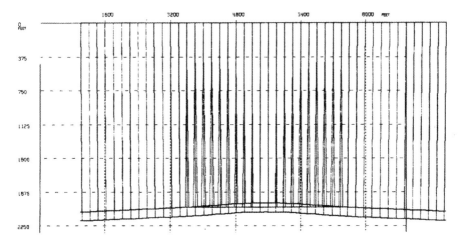

Fig. 6.24 Ray tracing of the model

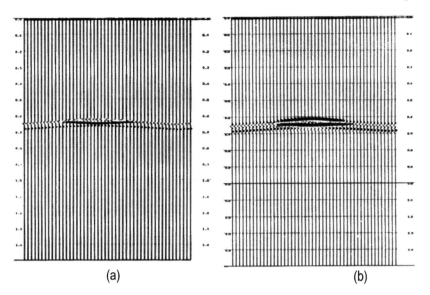

Fig. 6.25 (**a**) Spike seismogram from the model and (**b**) wavelet seismogram from the model

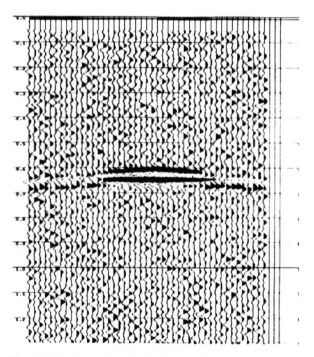

Fig. 6.26 Random noise added to the wavelet seismogram

Determine the error between the model section and the seismic section. If the error exceeds the maximum allowed, modify the earth model parameters and recalculate the model section. Test and repeat (iterate) as necessary to reach agreement with the seismic section. Keep parameters confined to the local physical limits.

Three Dimensional Modeling

Two-dimensional assumptions limit the interpreter's analysis. The subsurface is a three-dimensional solid so three-dimensional modeling would be more appropriate to solving many problems encountered in 3-D data. Some 3-D modeling is now performed, and as computer technology develops, these techniques will improve in effect and reduce in cost.

Conclusions

Seismic sections output from seismic data processing can be thought of as seismic or acoustic models of the subsurface. The modeling techniques described here can be used to test assumptions about the geologic structure and stratigraphy that produced the acoustic model. Modeling thus provides a way to accurately define subsurface geologic structures and stratigraphy, which is the objective of seismic exploration. Physical constraints and the real data must be considered in all modeling.

Modeling forces a closer look at mechanisms causing seismic distortion at a cost is much less than drilling to test an interpretation. All in all, it is the best method available to geophysicists for providing insights required for data interpretation.

Tomography

Most people are familiar with "CAT Scans" and PET Scans. CAT is an acronym for *computer assisted tomography* and PET is an acronym for *positron emission tomography*. Both are used as non-invasive tools to investigate internal tissues in the human body, usually to obtain information about tumors. CAT Scans use X-rays as an energy source with many different configurations between source and receiver. Computer programs are used to analyze the recordings of X-ray reception to provide a picture of the body or some part of it. In PET Scans the patient swallows radioactive material that emits positrons (positively-charged particles the size of electrons). A very large number of positron detectors are arranged in a ring through which the patient passes. Analysis of variation in positron reception provides a picture of the part of the body scanned.

The principles of tomography are applicable to seismology, as well. For example, earthquake seismologists applied tomographic methods to produce a velocity model for the earth's mantle. Data from about two million earthquake travel times were used. Interest in applying tomography to exploration seismology is increasing.

Seismic tomography is a type of *inverse modeling* or *inversion.* There are two types of modeling: forward and inverse. *Forward modeling* makes use of earth models, seismic data acquisition parameters, and theoretical models of physical processes to generate synthetic data that match, within some maximum error, actual seismic data. *Inversion* is similar to forward modeling in that it uses data acquisition parameters and theoretical models of physical processes to produce synthetic data. However, the objective of inversion is to produce a model of the subsurface that predicts the observed data.

Only a few publications describing seismic tomography field experiments exist. Weatherby was one of the early American geophysicists to propose estimating seismic velocities along ray paths between boreholes. Bois, et al. (1972) not only explained this can be done but also demonstrated their results with real field data. Butler and Curro (1981), along with many other geophysicists, have discussed procedures for recording cross-hole seismic data.

In recent years, geophysicists have successfully used seismic tomography to image velocity variations of the earth. In so doing the accuracy of depth conversion, depth migration, and other applications are enhance. This will be discussed later in this section.

Types of Seismic Tomography

There are two types of seismic tomography: *reflection* and *transmission.* Reflection tomography involves seismic waves that propagate from the surface to a subsurface reflecting marker and back to the surface. *Transmission tomography*, involves seismic energy that has traveled *through* the subsurface without reflection.

Reflection tomography requires ray-tracing computations, for which it is important to define reflecting markers or boundaries. As a result reflection tomography is difficult to model.

Gulf Oil Company geophysicists wrote, in 1980, that seismic velocities can be successfully estimated from seismic reflection times using travel-time tomography. Scientists at Amoco Research Company and at the California Institute of Technology have demonstrated this method. The velocities obtained can be used to enhance seismic imaging processes such as migration and depth conversion.

Transmission tomography requires placing the source in a borehole and the receivers at the surface, or vice versa. Alternatively, the source may be in one borehole and the receivers in another.

Measurements of first arrivals in a vertical seismic profile (VSP. See Chap. 8) may be used in borehole-to-surface tomography. In VSP, seismic energy is injected into the ground and the resulting waves are recorded by a borehole geophone at various depths in the borehole. In another method of transmission tomography, energy is injected at depth in the borehole, using a downhole source, and is recorded at the surface (reverse VSP).

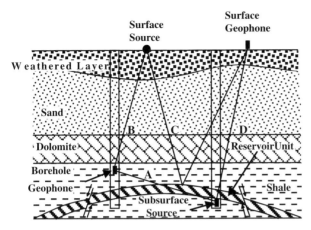

Fig. 6.27 Raypaths between surface and subsurface source and receiver positions

In borehole-to-borehole measurements, both sources and receivers are placed in-side boreholes. Seismic energy "illuminates" the region between boreholes. Figure 6.27 shows types of raypaths between surface and subsurface source and receiver positions. Ray path B is a downward transmission, ray path D is an upward transmission, and ray path A is a lateral transmission. Note that ray path A does not travel through the weathered layer resulting in little loss of the high frequency components of energy caused by scattering and absorption. Ray path C is the normal type for surface reflection seismic surveys. Since it is the longest ray path and passes through the weathered layer twice, the greatest amount of high frequency attenuation results. It is, however, the only ray path in Fig. 6.27 that provides reflection information from the reservoir.

Travel Time Tomography Procedure

Both reflection and transmission type tomography, involve the following:

- Determination of actual seismic traveltimes.
- Ray trace modeling of energy travel paths.
- Solution of traveltime equations to implement "traveltime inversion". So called because it is used to produce a velocity model that best fits the observed data.

Determination of Travel Time

Travel times can be picked from common source (shot) records, common midpoint (CMP) records, or slant stack (Radon-transformed) records. Picking is facilitated by using CMP records output from deconvolution since the reflection wavelets are zero-phase.

Ray Tracing

Tomography is based on the assumption that the energy travels from source to receiver along a particular ray path. Ray tracing models the energy propagation through a medium by solving equations applied to a velocity model, a set of reflecting boundaries, and receiver pairs (Aki and Richards, 1980). The solution to the two-point boundary for the ray equation is found by shooting upward from the reflecting surface. Ray path modeling can do this.

Travel Time Inversion or Tomographic Inversion

A layered-media model, normally subdivided into constant-velocity cells, (see Fig. 6.28) is used in tomographic inversion. Unless all cells have the same velocity, bent ray paths will result. Ray-bending methods are generally needed.

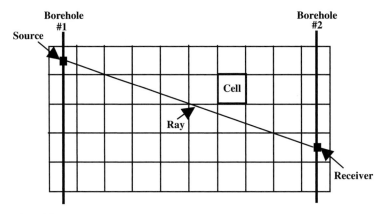

Fig. 6.28 Layered media model

Times in each cell are simply the length of the path in the cell divided by the cell velocity. However it is computationally more efficient to use slowness (the reciprocal of velocity or $1/V$)

Although there are complicating factors such as many of the n x m cells not having any ray components in them a least-squares solution of the system eventually results by iteration. Once the actual traveltimes have been picked and rays have been traced, a large system of traveltime equations is solved for the unknown slowness equation.

Transmission Tomography

Transmission tomography can be done as either borehole-to-borehole or borehole-to-surface. Both are shown in Fig. 6.29. On the left is the borehole-to-borehole type

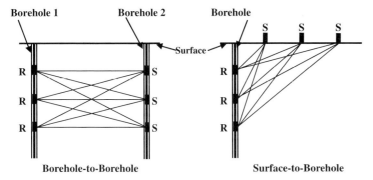

Fig. 6.29 Transmission tomography geometry

with sources placed below the surface in one borehole, and receivers placed in another hole. An advantage of this arrangement is that the signals recorded are richer in high-frequency components than in the surface-to- borehole arrangement. This is because there is no travel through the near-surface weathered formations where the attenuation of higher frequencies is most pronounced. Surface-to-borehole geometry is typical in VSP surveys.

Models used in both reflection and transmission tomography will be discussed before any real data examples are presented.

Reflection Travel Time Tomography Model

Figure 6.30 illustrates the reflection tomography geometry used by Bishop et al. (1985) to demonstrate the finite difference approach.

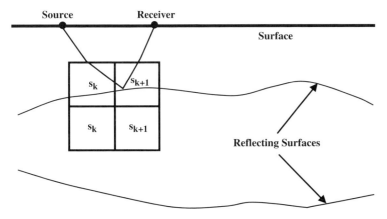

Fig. 6.30 Reflection tomography geometry

The characteristics of this approach are:

1. The subsurface model is divided into cells that have constant slowness.
2. Curved rays are traced from the source to the reflecting horizon and back to the surface.
3. The reflecting horizons are known.

Figure 6.31 illustrates an iterative procedure used by Amoco Research scientists (Bording et al., 1987) to image a velocity field. This procedure combines the ability of tomography to extract the subsurface velocity field with that of migration to image the subsurface interfaces. This is called *iterative tomographic migration.*

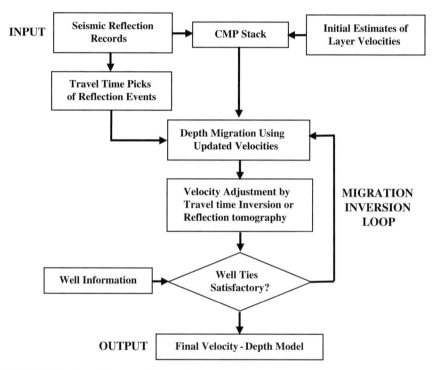

Fig. 6.31 Iterative reflection and migration tomography

Iterative Tomographic Migration Procedure

The geological model of Fig. 6.32(a), which also shows velocities of the layers, was used to demonstrate this approach. The initial velocity model has flat, constant-velocity layers. The velocities on the left side of the model section were chosen to be correct. Figure 6.32(b) shows a CMP stack generated from this velocity model.

The velocity model was used to generate the depth-migrated section of Fig. 6.32(a). The migration of the dome's left side is a good reconstruction but not of the right side of the salt dome or deeper markers of the model.

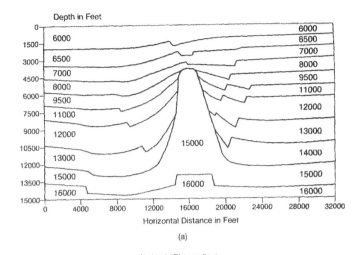

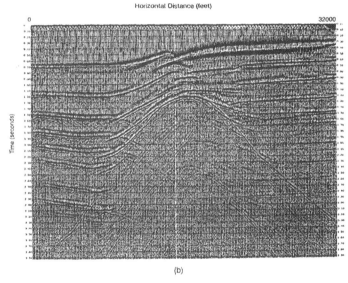

Fig. 6.32 (**a**) Geological model used in reflection tomography example with layer velocities shown. (**b**) CMP stack using flat layer velocity model

To improve the flat-layer migration results, an accurate velocity tomogram was needed to provide refined velocities. For this purpose, a set of well-spaced common shot records were chosen (see Fig. 6.33b). Travel times of ten events within each of nine records were picked, resulting in a total of 4,468 travel-times being available.

Ray Tracing for the Model

The model is 40 cells wide and 80 cells deep. Each cell has dimensions of 400 × 400 ft. Fig. 6.34 shows the rays traced on three different reflectors.

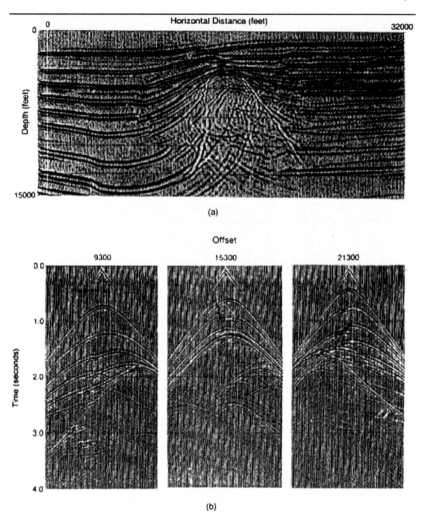

Fig. 6.33 Iterative tomographic migration (**a**) initial depth migration using flat layer velocities. (**b**) finite difference common source gathers for offsets of 9,500, 15,300, and 21,300 ft (Copyright © 1987, Blackwell Scientific Publications, Ltd., from Bording et al., "Applications of seismic travel—time tomography," *Geophysics Journal International,* vol. 90, 1987)

The first step in tomographic inversion was to determine the difference, ΔT, between the estimated travel times from the data and the computed travel times from the model. An equation (not shown here) was solved for the slowness deviation ΔS, where ΔS is the difference between the data and initial model response. Only cells that are illuminated contribute significantly to the inversion. Consequently, sparse iterative solutions were employed.

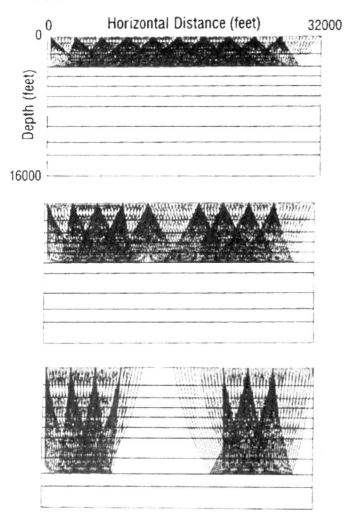

Fig. 6.34 Ray paths traced to three reflectors based on model of Fig. 6.35(b)

Next, the slowness deviations, ΔS, are added to the original slowness model to update the slowness solution. A filter is usually applied to the output velocity model to smooth the solution.

These two steps can be repeated until the variation in ΔS reduces to zero, resulting in an optimum solution. Hence, the tomographic inversion process is completed with satisfactory results;

The final velocities determined from the depth model are used to perform the final stack. See Fig. 6.35. This new CMP stack section is then depth migrated with the tomographically derived velocities. As can be seen in Fig. 6.36, the result is a much-improved image of the salt dome flanks compared to Fig. 6.36(a).

0   Horizontal Distance (feet)   32000

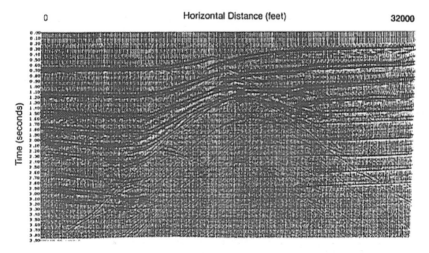

Fig. 6.35 CMP stack using tomographically derived velocities

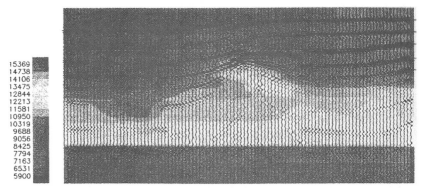

Fig. 6.36 Depth migration using tomographically derived velocities overlain by computed tomogram of velocity range from 6000 ft/s to 16,000 ft/s, (after Bording et al., 1987)

Cross-Borehole Tomography Model

In reflection seismology, significant high-frequency attenuation often occurs near the surface because of the heterogeneity of weathered layers. This problem is much reduced in cross-borehole seismic methods because both source and receivers are below the weathering.

A seismic time section or map is distorted. What appears to be a small separation between horizons may actually be large because the interval velocity is high. Conversely, an apparently large separation may be, in fact, small because the interval velocity is low. Time-to-depth conversion eliminates this type of distortion. This however, requires velocity control.

Sonic logs provide velocity control at the borehole and velocities between boreholes can be interpolated. Alternatively, and preferably, velocity between boreholes can be estimated from seismic travel time measurements made in a cross-borehole survey.

Note that tomography can not only provide velocity information for depth conversion and depth migration but can also be used to delineate reservoir boundaries and provide data for enhanced oil-recovery projects.

Near-Surface Model

Another application of transmission tomography is to model the near-surface in an area. This approach was documented by Lines and La Fehr (1989) in a study based on a data set from Amoco's Denver Region. This study may help in getting better record quality in future field data acquisition.

A classic paper by Wiggins et al. (1976) approached residual statics analysis as a general linear inverse problem. In this paper it was assumed that "elevation statics" and NMO corrections had been applied to the data traces. Thus, data had been adjusted for time differences caused by variations in shot and receiver elevations plus weathering variations that are determined by uphole times. These assumptions were based on the following:

1. Large apparent lateral variations in the velocities are caused by near-surface formation rather than to velocity changes at depth.
2. Lateral averaging of residual NMO tends to stabilize the statics solution.

Multiplicity of the data is an important factor in the accuracy of the near-surface model obtained. The higher the multiplicity, the larger the statistical sample and the closer the statics solution will be. A minimum of three fold stack is required for reasonable results. Figure 6.37 illustrates this method by showing data before and after applying the statics derived from tomography. It was known that the marker at 2.0 s is almost flat, as it is after applying the statics derived from tomography.

Error Criterion and Forward Model

Inversion requires a forward model with a known response. In the case of a geological model, its geophysical response must be known. An optimization algorithm is also needed so that parameters of the forward model adjusted.

These choices were discussed by Treitel in his 1989 paper. The choice of the forward model and the optimization algorithm greatly affect the quality of the inversion. A very important question is how to minimize the difference between the observed and modeled? Choices include least- square error, the least absolute deviation, and mini-max methods. Treitel proposed conducting a sensitivity analysis as an important step in deciding among these. For example, if a in a set of parameters that

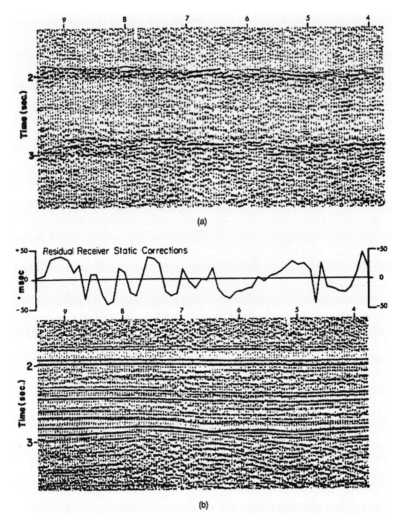

Fig. 6.37 Residual statics tomography. (**a**) 12-fold stack of 48 trace records. Elevation statics were applied. One cable length spans 4 intervals. (**b**) same section as above after surface-consitent statics were applied. Residual statics exceed 50 ms in some portions of the line. (After WesternGeco)

have been extracted using an inversion procedure, it is found that changing a particular parameter by 10, 15, and 20% produces no change., the model is obviously not sensitive to this parameter.

The choice of the *initial* model is extremely important. A close guess can result in a good inversion.

Figure 6.38 illustrates this point for a pre-stack model. Figure 6.38(a) shows a model of common source gather for a four-layer case. This is what the inversion is to duplicate. Figure 6.38(b) shows the change in the initial model after one iteration of the velocity inversion. This indicates that the first-guess model has no match with

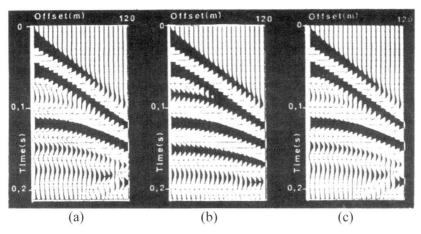

Fig. 6.38 Seismograms for model Inversion, (courtesy of Society of Exploration Geophysicists, adapted from Treitel, 1989)

the model in (a). Figure 6.38(c) shows that, after 10 iterations, a good match with the model in (a) is obtained.

Figure 6.39 (a) is the starting velocity model used in the velocity inversion for Fig. 6.41. A good guess results in a large change toward the correct velocity model, after only one iteration, as shown in Fig. 6.39(b). Figure 6.39(d) shows that a very good match to the actual velocity model is obtained after 125.

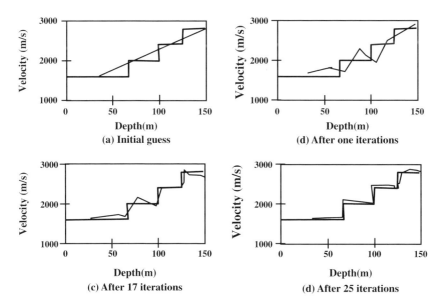

Fig. 6.39 Progress of velocity inversion – good initial guess of model (after Treitel, 1989)

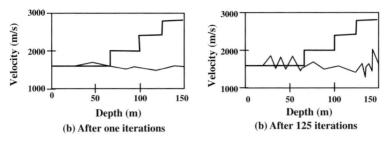

Fig. 6.40 Progress of velocity inversion – bad initial guess of model (after Treitel, 1989)

Figure 6.40 shows what happens when starting with a bad initial guess of the velocity model, Here, after the same number of iterations, 125, a bad velocity model results.

Seismic Tomography and Reservoir Properties

Much better descriptions of existing reservoirs are needed to improve recovery from them. In particular, knowledge of their internal structure and fluid flow through them is needed. Considerable information about reservoir properties can be obtained from cores, cuttings, well logs, and well testing. These data, however, are sparse. Improve the sampling of the subsurface is needed to get a better description of the reservoir. Drilling enough wells to get the required data to accomplish this is out of the question. Some other method of acquiring these data must be used. Seismic borehole measurements and tomography may provide the data needed to increase the understanding of reservoir properties.

3-D surveys and extensive VSP surveys better sample reservoirs than in the past. Reservoir geophysics, especially seismic reservoir monitoring, are vital to the understanding of reservoir characteristics. Borehole seismic surveys are of immense help in relating wave propagation to the reservoir structure and fluid flow behavior to the seismic velocity response. The better methods now available to acquire and process the seismic response will continue to revolutionize reservoir engineering, oil recovery, and enhanced oil-recovery methods.

Since reservoir rock is heterogeneous, reservoir properties need to be obtained in their spatial and temporal variations, both laterally and vertically. The properties needed include:

- Mineralogy.
- Rock properties, porosity, permeability, compressibility, and saturations.
- Fluid properties—chemistry, viscosity, compressibility, and wettability.
- Environmental factors—temperature, stress, and pore pressure.

A more complete understanding of the relationships between reservoir properties, fluid flow, and seismic wave propagation, is needed. This is to assure that the application of existing methods and future developments of hardware to acquire more measurements and software to expedite the results will not be hindered. It is also requisite that the geophysical and engineering communities should work together in this endeavor. This should include efforts to understand each other's terminology, to appreciate the complexity of the methods geophysicists are developing, and to understand more about their applications and limitations.

More and more reservoirs are being classified as heterogeneous. Thus, it is of primary importance to use tomography to describe reservoir heterogeneity. Monitoring recovery of hydrocarbon in place is an important application but recovery plans may be re-evaluated as production proceeds.

Laboratory work has been used to relate acoustic velocity and seismic response to petrophysical properties of reservoir rock. It may be possible to develop a relationship between dispersion in seismic measurements to hydraulic permeability.

Examples of such laboratory work are shown in Figs. 6.41, 6.42, 6.43 and 6.44. The relationship between porosity and clay content is shown in Fig. 6.41. A total of 75 shaly sandstones are used to show the effect of both P- and S-velocities.

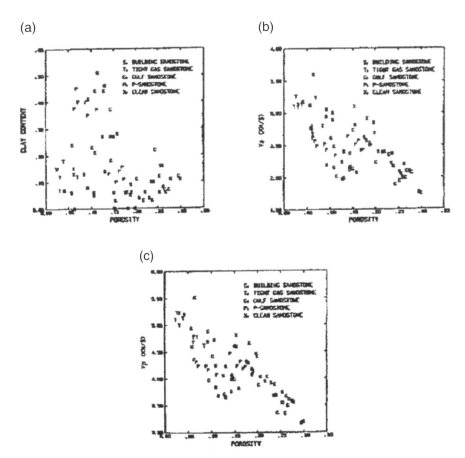

Fig. 6.41 Effect of porosity and clay content on velocity. (**a**) clay content versus porosity, (**b**) compressional velocities versus porosity, (**c**) shear velocity versus porosity

Fig. 6.42 Effect of
temperature on velocity (after
Nur, 1989 courtesy SEG)

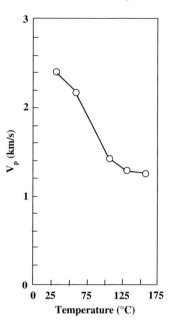

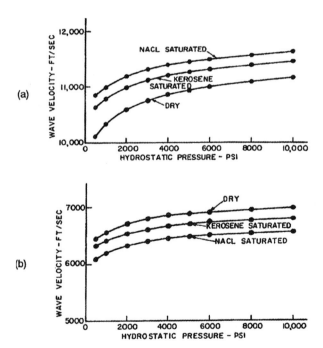

Fig. 6.43 Effect of saturation and pressure on boise sandstone group velocities. (**a**) P-wave.
(**b**) S-wave (after King, 1966 courtesy SEG)

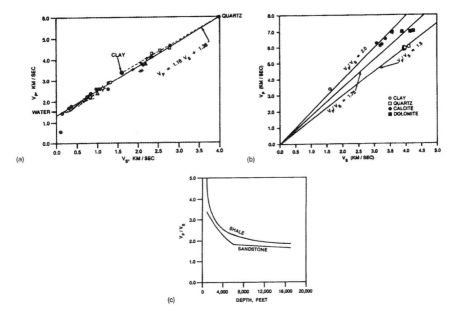

Fig. 6.44 V_p versus V_s. (**a**) some minerals. (**b**) mudrocks. (**c**) V_p/V_s computed as a function of depth (after Castagna, 1984 courtesy SEG)

Figure 6.42 shows the effect of temperature on velocity. Figure 6.43 shows the effects of saturation and pressure on velocity. Figure 6.44 shows the relationships between P- and S-wave velocities for various minerals, as a function of depth for selected Gulf Coast formations.

Amplitude Versus Offset Analysis

The amplitude of a reflected seismic signal normally decreases with the increase of the distance between source and receiver. This decrease is related to the dependence of reflectivity on the angle at which the seismic wave strikes the interface, spreading, absorption, near surface effects, multiples, geophone planting, geophone arrays and instrumentation.

In certain depositional environments, the amplitude variation can also be an important clue to the lithology or to the presence of hydrocarbons. Figure 6.45 illustrates three classes of such compressional wave amplitude anomalies. In class 1, the reflection coefficient is positive at normal incidence and becoming negative at large angles of incidence. In class 2, the reflection coefficient is about 0 for normal incidence and become increasingly negative. In class 3 the reflection coefficient is negative at normal incidence and becomes more negative as the angle of incidence increases. Class 3 is a typical bright spot anomaly (gas sands).

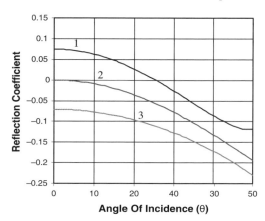

These amplitude anomalies are masked in the common midpoint stack (CMP), as every trace of the stack section represents an over-all average of offsets in the common midpoint gather. AVO analysis is designed to retrieve the variation in amplitude with angle of incidence by conducting the analysis on the normal moveout corrected gathers *before* stack.

Review of AVO Development

Interpretation of a seismic section is restricted to the zero-offset model. Accordingly, an incident plane wavefront of amplitude A_0 on a horizontal interface will produce a reflected plane wavefront of amplitude of A_1. The ratio of A_1 to A_0 is defined as the reflection coefficient (R) of this interface, which depends upon the difference between acoustic impedances on each side of the interface. (Acoustic impedance is the product of density and velocity.)

In the case of CMP gathers, however, angles of incidence other than normal must be considered. In this case, an incident P-wave gives rise to a reflected and a transmitted S-wave as well as reflected and transmitted P-waves. So, for non-normal incidence the amplitude of a seismic reflection depends on an additional rock property – shear wave velocity, V_s.

Amplitude Versus Angle of Incidence

Zoeppritz derived a set of equations for plane wave reflection and transmission coefficients as functions of the angle of incidence, density, and velocities. These equations are not difficult to solve but require P and S velocities and densities on each side of an interface as well as angle of incidence for each calculation.

Shuey developed a simplification the Zoeppritz equations that can be more readily used to analyze the offset dependence of event amplitudes.

Confirmation of Shuey's Approximation

For a given interface, when acoustic and elastic properties are given, both the Zoeppritz and Shuey equations can be applied to obtain a relationship between reflection coefficient and angle of incidence (in degrees). The two equations give the same results up to 10° angle of incidence, and they do not differ significantly up to 45°.

The example of Fig. 6.46 is for a typical Gulf Coast gas sand. The parameters used are:

Layer 1 – $V_p = 10,000$ fps, $\rho = 2.2$ gm/cm^3,

$\qquad \sigma = 0.20$

Layer 2 – $V_p = 12,000$ fps, $\rho = 1.95$ gm/cm^3

$\qquad \sigma = 0.40$

where V_p is P = wave velocity, ρ is density, and σ is Poisson's ratio.

One can see the increase of reflection coefficient (amplitude) with the increase of angle of incidence or offset.

Figure 6.47 is a gas sand. Parameters are:

Layer 1 – $V_p = 7,570$ fps, $\rho = 2.15$ gm/cm^3,

$\qquad \sigma = 0.40$

Layer 2 – $V_p = 6,400$ fps, $\rho = 1.95$ gm/cm^3

$\qquad \sigma = 0.10$

Again, the fit between the Shuey and Zoeppritz calculations is quite good.

Figure 6.48 illustrates a decrease of amplitude with increase in angle of incidence. This is a typical *dim* spot as observed in carbonate rocks. The curve was

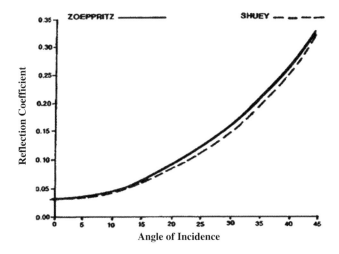

Fig. 6.46 Reflection coefficient comparison – typical gulf coast sand

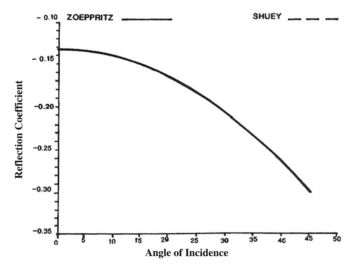

Fig. 6.47 Gas sand

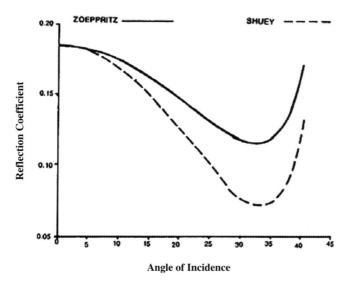

Fig. 6.48 Carbonate dim spot

derived from the modeling of the Austin Chalk formation in the Texas Gulf Coast.
Parameters are:

Layer 1 – $V_p = 8,000$ fps, $\rho = 2.15$ gm/cm^3,
$\quad\quad\quad \sigma = 0.40$
Layer 2 – $V_p = 11,360$ fps, $\rho = 2.2$ gm/cm^3
$\quad\quad\quad \sigma = 0.30$

Notice the departure of the curves from the two methods of computing the angle of incidence. Yet, both show the same trend of the relationship between reflection coefficient and the angle of incidence.

Zoeppritz equations give the complete solution that relates change in amplitude with the angle of incidence. The other approximations such as Shuey's are acceptable to a certain extent for most lithologies. Other approximations are suitable for some localized and specific areas.

AVO Analysis Methods

A CMP gather orders traces by offset, not angle of incidence but what is needed is relative P-wave amplitude as a function of angle of incidence, not offset. The direct analysis is actually AVA (Amplitude Versus Angle), not AVO. Consequently, the input CMP data must be converted from offset-ordered to angle of incidence-ordered.

Constant-Angle Stack

Traces recorded at fixed offsets can be transformed to traces at fixed (or a limited range) of angles of incidence. This allows observation of amplitude variation with reflection angle. Figure 6.49 shows the variation of reflection angle with depth for a fixed offset. Figure 6.50 shows that traces at different offsets traces may have the same reflection angle.

The transformation is made via constant angle stack gathers. Each angle trace is generated by partial stacking traces in an NMO corrected CMP gather. The extent of partial stacking is by an angle range width or window beam. The annotated angle represents the central angle of the range.

Figure 6.51 illustrates this approach. The top part of the Figure shows the angle traces produced from CMP gathers 310, 314, and 318, that have been corrected for normal moveout. The constant angle range is from $2°$ to $30°$. Bar graphs representing the amplitude variation with the angle of incidence are plotted below each set of constant angle stack gathers. As can be seen, amplitudes within the window of investigation (1.6–1.7 s) increase as angles of incidence increase on all three CMPs, but it is most pronounced on CMP 318. A curve of RMS or maximum amplitude can be plotted to define this anomaly.

AVO Attributes and Displays

Various parameters can be displayed on sections in a manner similar to the conventional stack. These displays assist analysis of AVO. One of these is a near-trace stack, which is generated by stacking only short offset traces selected from each

DISTANCE

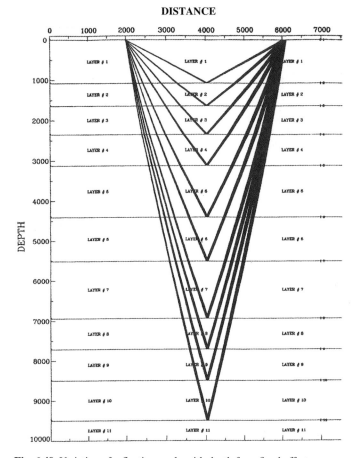

Fig. 6.49 Variation of reflection angle with depth for a fixed offset

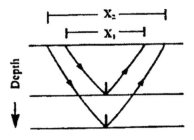

Fig. 6.50 Same reflection
angle at different offsets

CMP gather, corrected for NMO. Stacking only longer offset traces selected from
each CMP gather, corrected for NMO, generates a far-trace stack. Figure 6.52 is a
near-trace stack and Fig. 6.53 is a far-trace stack. Note the amplitude anomaly on
the far-trace stack at around 1.6–1.7 s that does not appear on the near-trace stack.

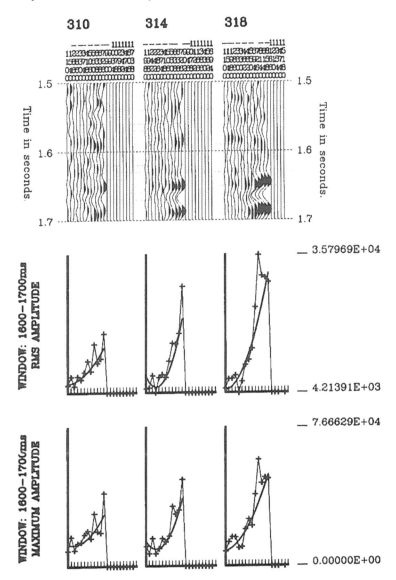

Fig. 6.51 Angle stacks generated from three CMP gathers and amplitude variation with angle

Other useful parameters can be generated from plots of amplitude versus $\sin^2\theta$. This is done for every sample time of the angle stacks. Linear regression is used to fit straight lines to the plotted data.

The constant A is called the *AVO Intercept* and B is called the *AVO Gradient*. Figure 6.54 shows a plot of amplitude versus $\sin^2\theta$ The intercept, A, is found by projecting straight line back to $\sin^2\theta = 0$. Note that if relative amplitude is plotted, the AVO intercept is actually the normal incidence P-wave reflection coefficient

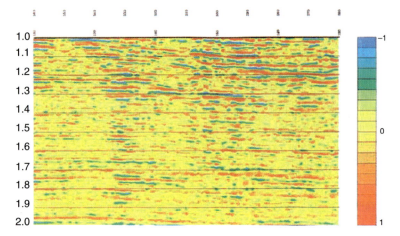

Fig. 6.52 Near trace stack

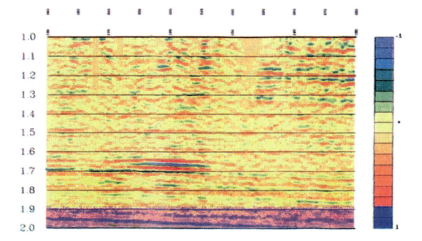

Fig. 6.53 Far trace stack

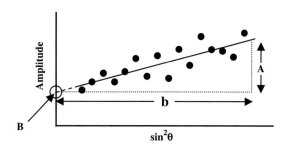

Fig. 6.54 Amplitude versus
$\sin^2 \theta$

since at normal incidence $\theta = 0$. Referring to Fig. 6.54, the gradient is found from $A = (a/b)$. The AVO intercept is often symbolized by R_0 instead of B and the AVO gradient by G instead of A. Since plots are made for every sample time of the angle stacks, values of A and B (or G and R_0) are defined for each sample time. These values can be plotted against time to produce a gradient and an intercept trace for each CMP used in the analysis. Combining the all the gradient traces produces a gradient section. Similarly, combining all the intercept traces produces an intercept section.

The constants A and B are functions of various elastic parameters. They can be combined to produce other meaningful data. For example, $(1/2)(A - B)$ can be use to produce what is called a *Pseudo-S Wave Section* and $(4/3)(A + B)$ can be used to produce a Poisson's Ratio Contrast Section.

Figure 6.55 is an AVO gradient section derived from the same CMPs as the near-trace and far-trace stacks of Figs. 6.52 and 6.53. The amplitude anomaly (bright spot) seen on the far-trace stack is also apparent here.

Figure 6.56 shows a part of a CMP stack section and, below it, displays of parameters derived from the CMPs: P-wave reflection (or AVO intercept) section, pseudo S-wave section, and Poisson's ratio contrast section.

Each trace of the P-wave reflection section shows the sequential variation of P-wave reflectivity. The pseudo S-wave section mimics what would be recorded in a zero-offset S-wave section but with travel time determined by P-wave velocities. Each trace of the Poisson's ratio section shows the change in Poisson's ratio with P-wave record time. An increase in the amplitude of a seismic event with the increase in offset from the source to receiver (related to increased angle of incidence) indicates a geological marker. In this case, the bright spot is associated with a gas sand reservoir. increased angle of incidence indicates a geological marker. In this case, the bright spot is associated with a gas sand reservoir.

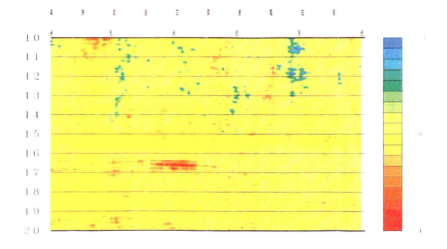

Fig. 6.55 AVO gradient stack

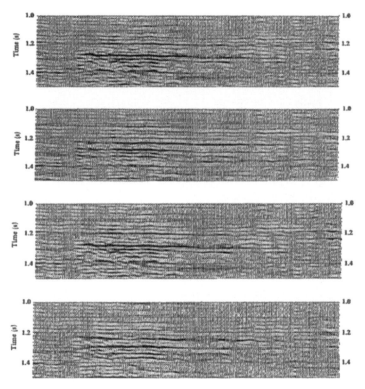

Fig. 6.56 From *top* to *bottom*: part of a CMP stack section showing a bright spot, P-wave reflection coefficient section, pseudo S-wave section, and poisson's ratio section

Data Processing

Traces used in AVO analysis require somewhat different processing. Figure 6.57 illustrates data processing flow designed to preserve and enhance the true amplitude of each trace within the CMP.

The first difference between conventional processing and processing data for AVO analysis is the geophone array correction. Figures 6.58 and 6.59 demonstrate the need for this correction.

Figure 6.58 shows 12 geophones planted in line on the ground over a distance of 220 ft. Figure 6.59 is a synthetic record showing signal arrivals at geophones 1–12, arranged from left to right. The time difference between geophone 1 and geophone 12 for the first reflector is about 10 ms. The individual geophone signals are summed together in ore trace on the extreme right. Note that the summed trace, recorded as the array response, lacks some of the high frequencies seen on the individual traces. Summing traces with time shifts from geophone to geophone within the array acts as a high frequency filter.

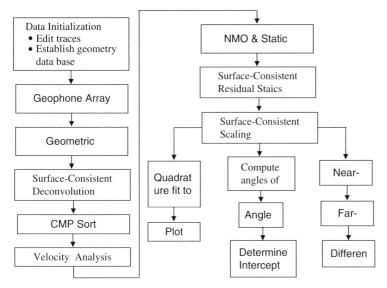

Fig. 6.57 Data processing flow chart for AVO analysis

Fig. 6.58 Geophone array correction

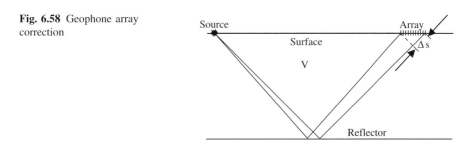

The differential time within the array from the first geophone to the last geophone decreases with depth, as the angle of incidence decreases with depth. It is critical to maintain the high-frequency component up shallow, especially if the data is recorded for shallow targets. Note that the geometrical spreading correction is applied but not a gain function. Gain is a statistical process and can destroy or greatly modify amplitude differences.

Conventional spiking or gapped deconvolution must also be avoided, as it is also a statistical process that tends to modify amplitude relationships. Surface consistent deconvolution only considers factors in the vicinity of sources and receivers. Signature deconvolution can be applied, if a valid source signature is available.

Residual statics analysis should also be surface-consistent. Conventional residual statics involve trace summing. Scaling is a critical step, and it should be done in a surface-consistent manner. Amplitude variations resulting from conditions near sources and receivers should be removed. Scale factors for each source and receiver on the line accomplish this purpose. However, these factors should be edited. Final surface-consistent scaling is achieved when all the amplitude values are approximately equal.

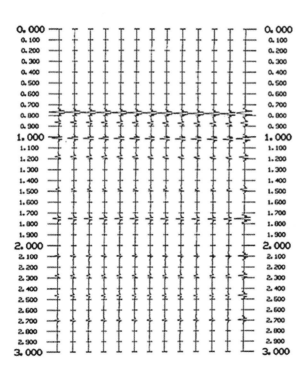

Fig. 6.59 Synthetic record for an array of 12 geophones

Processing operations that must be avoided in order to preserve the amplitude versus offset relation include multi-channel operations such as:

- Mixing traces.
- Trace-to-trace scaling with small windows.
- F-K operations.
- Deconvolution derived from trace summation.

These operations remove the significance of true amplitude.

Advantages of AVO

AVO provides verification of direct hydrocarbon indicators such as bright spots in gas sands and dim spots in carbonate reservoirs, as well as related amplitude anomalies. AVO analysis based on NMO-corrected CMP gathers is a two-dimensional analysis compared to a stacked trace, which is a one-dimensional analysis.

Amplitude variation with the angle of incidence is another tool used to confirm observed anomalies. It also allows estimation of Poisson's ratio, for a given rock.

This, along with velocity information, can be used to determine elastic properties of subsurface rocks.

Applications of AVO

Applications of AVO analysis include:

- Reservoir Boundary Definition
- Identifying "Look-Alike" Anomalies
- Predicting High Pressure Gas Zones

These are briefly discussed below.

Reservoir Boundary Definition

Estimating the hydrocarbons in place is one of the more difficult tasks for a reservoir engineer or geologist. This is because there are so many variables involved, especially in the exploration stage. Among these variables is the areal extent of the reservoir. It is important, in the development stage of a field, to define the boundary of the reservoir so that the optimum spacing between development wells can be determined.

Given adequate coverage of sufficient density, seismic data that have been processed for AVO analysis, the boundary of the reservoir can be defined with a great deal of accuracy. A 3-D seismic survey is preferred as the cost of the field data acquisition and data processing become economical.

Identifying "Look-Alike" Anomalies

There are AVO analyses of anomalies that have been confirmed and documented by drilling and successful completion. These can be used to locate look-alike anomalies in other geological provinces that have the same or very similar geological and lithological settings. The AVO analysis should always be integrated with other geological and engineering information.

Predicting High Pressure Gas Zones

Bright spot anomalies are usually associated with gas sand formations or lenses. Some of these lenses have been found to have abnormally high pressure at relatively shallow depths. Where such geological conditions are known to be present in an area, an AVO analysis of seismic data may confirm the anomaly. This warns the drilling engineer to take precautions for the existence of a high-pressure zone and to modify his drilling program before the bit hits the zone.

VSP Data Interpretation

Vertical seismic profiling is applied in two principal areas:

- Exploration, and
- Reservoir engineering and drilling.

  Exploration applications include:

- Identifying seismic reflectors.
- Comparing VSP and synthetic.
- Studying seismic amplitudes.
- Determining rock properties such as seismic wave attenuation.
- Investigating thin bed stratigraphy.

  Reservoir engineering and drilling applications include:

- Predicting depths of seismic reflectors.
- Predicting rock conditions ahead of the bit.
- Defining reservoir boundaries.
- Locating faults.
- Monitoring secondary recovery processes.
- Describing reservoirs via seismic tomography.
- Predicting high-pressure zones ahead of the bit.
- Detecting man-made fractures.

  Some of these topics are discussed, below.

Exploration Applications

A major goal of seismic interpretation is to relate surface-acquired reflection seismic data to subsurface stratigraphy and depositional facies. Achieving this goal is facilitated by using good quality VSP data to define the depths of the reflectors from which primary reflection arise. Thus, VSP interpretation complements and enhances interpretation of a surface-acquired reflection data.

VSP data, with a high signal-to-noise ratio can be used to resolve questions such as:

1. What is the nature of the boundaries at which seismic reflections occur?
2. Which rock boundaries can be seen with seismic data and which cannot?
3. Are synthetic seismograms made from well log data reliable means to identify primary and multiple reflections?

In some wells the stratigraphic and lithological conditions that create seismic reflections can be identified. Figure 6.60 shows VSP data recorded in such a well. Four upgoing primary reflections are shown by the lineup of black peaks labeled a, b, c, d. The subsurface depth of the interface(s) that generated each reflection can be defined by extrapolating the apexes of the black peaks downward until they intersect the first break loci of the downgoing compressional event. These depths

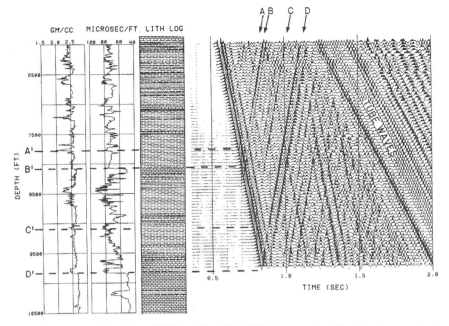

Fig. 6.60 An example of the reliability with which VSP data can often identify primary seismic reflectors.

are labeled A^I, B^I, C^I, D^I. These are raw field data. No processing has been done other than a numerical AGC function has been applied to equalize all amplitudes.

Figure 6.61 illustrates the tying of a VSP seismogram to the surface-acquired seismic data.

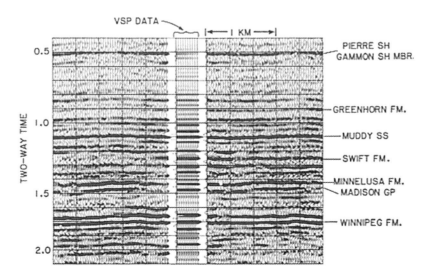

Fig. 6.61 Comparison of seismic data with VSP

Synthetic seismogram are usually the primary tool used to correlate subsurface stratigraphy and surface-measured seismic data. Vertical seismic profiling can be produced using the same type of source, a similar geophone, and the same instrumentation used to record the surface seismic data Synthetic seismograms, by contrast, are only *mathematical* representations of seismic measurements that can only approximate these aspects of the total seismic-recording process.

Figure 6.62 is a comparison of surface seismic data crossing VSP well "P" and well "Z" with a synthetic seismogram.

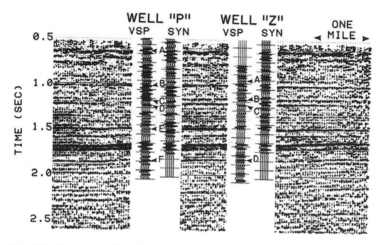

Fig. 6.62 Comparison of surface seismic data crossing VSP study wells "P" and "Z" with synthetic seismograms and VSP data recorded in the wells. The lettered arrowheads show where the VSP data are a better match to the surface data than are the synthetic seismogram data

One output of VSP data processing is a plot of acoustic impedance, the product of density and interval velocity, versus depth. Since change in densities of sedimentary rocks is much smaller than that of interval velocities, the change in acoustic impedance depends mostly on the change in interval velocity. If density is calculated from Gardner's equation, it can be removed from the acoustic impedance allowing interval velocity versus depth to be plotted, Interval velocities can also be used in engineering applications as shown in Figure 6.63 illustrates variation of acoustic impedance with depth derived from a VSP survey.

Reservoir Engineering and Drilling Applications

Determining drilling depth to key seismic markers is a valuable tool in drilling. Predicting these depths is a common practice in oil and gas exploration. While many geophysicists are able to make accurate depth estimates from surface determinations of seismic velocities, estimates are more difficult and less accurate when in a wildcat

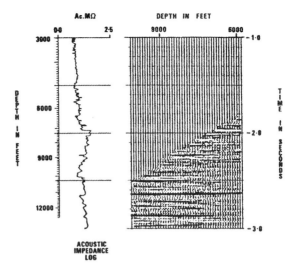

Fig. 6.63 Acoustic impedance versus depth display

area where there is little drilling history or where seismic recording quality is poor. It is in areas where seismic reflection is poor that using VSP to predict reflector depth is most beneficial.

A demonstration of this application is shown in Fig. 6.64. Since the downgoing wave trend intersects the upgoing wave trend at 9,850 ft depth the top of the reflector (A) is at this depth.

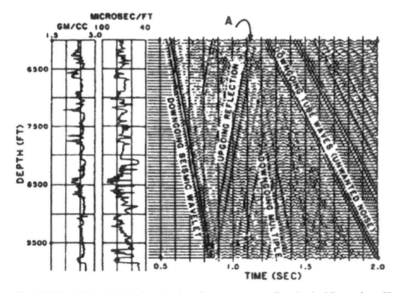

Fig. 6.64 Predicting depth of a seismic reflector (courtesy Geophysical Press, from Hardage, B.A.: "Vertical Seismic Profiling, Part A: Principles," 1983)

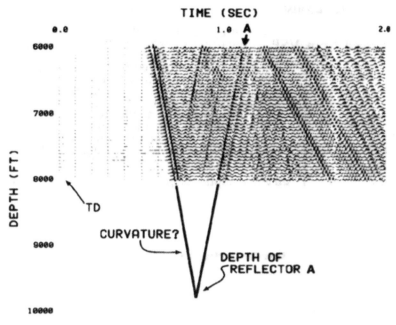

Fig. 6.65 Looking ahead of the bit

Figure 6.65 illustrates a way that VSP data has been used to predict the distance from the drill bit to a deeper formation. Assume that the well has been drilled to 8,000 ft. Further assume that VSP data are recorded far enough above 8000 ft for deep reflection events to be seen and interpreted. Data recorded from the bottom of the hole to about 2,000 ft above bottom, with constant depth increment should satisfy the latter assumption.

What is required is how far below the current drilling depth Reflector A is. To estimate this, extrapolate the downgoing first arrival wavelet trend below 8,000 ft with the same slope as the recorded data intervals from 6,000 to 8, 000 ft. Similarly extrapolate event A below 8000 ft. The two line extensions will intersect at the depth of Reflector A.

VSP can also be used to predict a variety of conditions beneath the drilling bit. For example, Stone (1982) and others have used VSP to predict interval velocity and depth beneath the bit. These two parameters are used to calculate pore pressure and porosity in well log analysis. There is a potential for calculating these important rock properties using VSP data. VSP data can be applied to amplitude correction, waveform and multiple extraction, deconvolution parameter design, and providing stratigraphic and lithologic identification of reflection events on surface-acquired seismic data. In addition VSP can be used to predict:

- Velocities ahead of the bit.
- Transit time versus depth ahead of the bit that can be used to calculate other petrophysical properties of the rock.

- Abnormal pressure zones ahead of the bit, allowing the drilling engineer to plan for pressure control.
- Porosity by using predicted transit time and knowing the transit times of other needed rock layers plus fluid in the borehole.

Stratigraphic Applications

A variation from conventional VSP, called offset VSP has shown to be useful in stratigraphic applications. Figure 6.66 illustrates offset VSP. First a conventional or "zero-offset" VSP record is made from the borehole geophone positioned at the depth of interest as in Fig. 6.66(a). Geophones are laid out on the surface, relatively closely spaced, over a distance L. A recording is made with the source at a small offset from the first geophone, as in Fig. 6.66(b). The source is moved a distance L from its first position and another record made (c). The process continues with the source moved a distance L between recordings (d).

A five-offset VSP implies five records are made from the surface geophones from five source positions spaced a distance L apart. Cramer (1988) showed the applicability of multi-offset VSPs to defining stratigraphy of the "D" Sand field in the Denver-Julesburg basin of Colorado. Well 34-3 established "D" Sand production in Wattenburg Field. (See Fig. 6.67). Three more wells were drilled but failed to determine the "D" sand's extent. A stratigraphic model suggested that the offset VSP technique could be used to delineate the extent of the "D" sand.

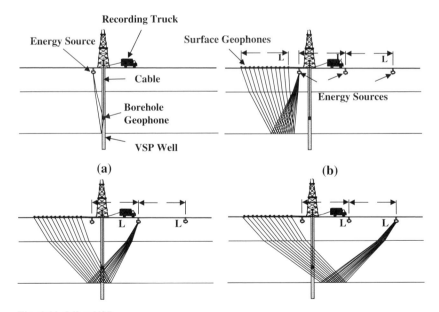

Fig. 6.66 Offset VSP

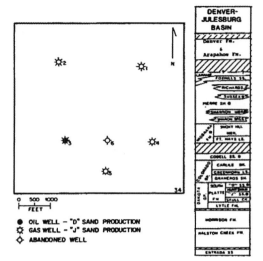

PLOT OF ACREAGE AND WELLS DRILLED IN SECTION 34

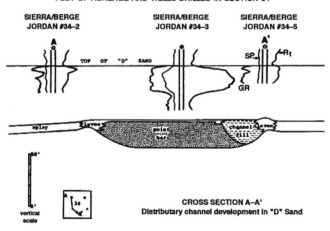

GEOLOGIC CROSS-SECTION

Fig. 6.67 "D" Sand field and geologic cross-section (Copyright © 1988, Society of Petroleum Engineers, from Cramer, P.M.: "Reservoir Development Using Offset VSP Techniques in the Denver-Julesburg Basin," *Journal of Petroleum Technology* (February 1988))

Consequently, a five-offset VSP was conducted in the discovery well. A zero-offset VSP was also run in this well to complete the delineation of the reservoir and to confirm the second location.

Survey Modeling

The geological model used to study VSP resolution and survey design is shown at the top of Fig. 6.68. The results of modeling synthetic offset VSP data correlated

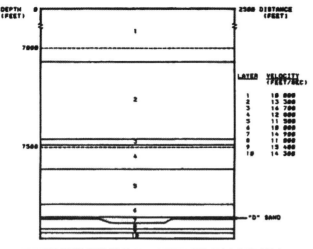

GEOLOGIC MODEL USED TO STUDY VSP RESOLUTION AND SURVEY DESIGN

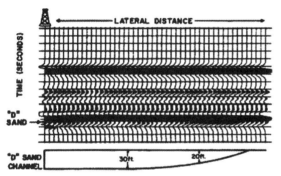

RESULT OF MODELING, SYNTHETIC OFFSET VSP DATA CORRELATED TO SAND "D" THICKNESS

Fig. 6.68 Survey modeling (Copyright © 1988, Society of Petroleum Engineers, from Cramer, P.N.: "Reservoir Development Using Offset VSP Techniques in the Denver-Julesburg Basin," *Journal of Petroleum Technology* (February 1988))

to Sand "D" thickness is shown at the bottom of Fig. 6.68. As indicated below the synthetic traces, sand thickness was allowed to change from 7 to 30 ft (2–9 m). A distinct character change can be seen that correlates with the known changes in the "D" Sand thickness. Modeling also demonstrated that reasonable survey parameters (i.e., source distances, geophone level spacing, number of levels, etc.) could be chosen to give a lateral profile length of 2,000 ft (600 m). From this model the decision was made to proceed with the VSP offset survey.

Despite some concerns Well 34-3 was selected as the site of the survey because:

- Its location allowed VSP data to be collected over most of the acreage held by the operator,
- The known existence of sand "D" in Well 34-3, would allow reliable correlation and calibration of the offset VSP to well logs.

Data Acquisition

Figure 6.68 shows both the original and modified plans for the VSP survey . Data were sampled at two ms. The vibrator sweep frequency was 10–80 Hz. The record length was 17 s with a listen time of 5 s. Six vibrators were used for the survey, two vibrators at each source point location. This required two runs of the well geophone into the hole, shooting from three offset locations on each run. According to the plan, the northwest, north, and northeast offsets were the first to be acquired. However, when the well geophone was pulled shallower than 6,300 ft no signal was observed. This depth coincided with the top of the cement behind casing. The solution was to adjust the source offset to get the desired lateral coverage of 2,000 ft (600 m) without raising the well geophone above the top of the cement.

The survey plan for the remaining far offsets was redesigned with four source points in each direction. In addition, the spacing between geophone levels was tightened to 50 ft (15 m) to provide closer spacing of reflection points in order to preserve the lateral resolution. Figure 6.69 shows a correlation of model data and zero offset VSP. Figure 6.70 shows the final offset VSP.

Data Interpretation

Figure 6.71 shows the final display of VSP data from each profile transformed to offset two-way time as in CMP seismic data. A check shot provided data for a time/depth chart to be computed from the zero-offset survey. The "D" sand was located from the zero-offset VSP, after comparing it to the model. The "D" sand response of each far offset was then identified and mapped away from the well (see Fig. 6.72).

Poor data recorded in the northwest, north, and northeast in uncemented casing complicated the interpretation. No reliable interpretation could be made through the zone. ; although Sand "D" on these offsets shows that it is disappearing midway in the record. It cannot be stated reliably that the edge of the buildup is where Sand "D" disappeared, but only where the sand extends to at least this point. It may extend further, but this cannot be determined from the data.

Subsequent Location

Well 34-7, located 1,650 ft (500 m) northeast of Well 34-3, was also completed in Sand "D". It had 19 ft (6 m) of oil pay, compared to 26 ft (8 m) of pay in Well 34-3. Another survey was planned to continue delineation of the reservoir extent and to locate a new development location, using Well 34-7. This survey was conducted in the open hole immediately after the well was logged and before casing was set. The survey, conducted according to the original plan was completed in 48 h.

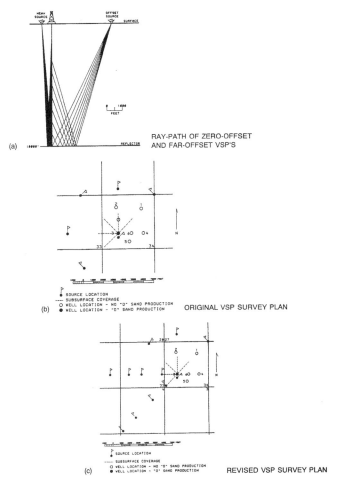

Fig. 6.69 Multi-offset VSP field plan (Copyright © 1988, Society of Petroleum Engineers, Cramer. P. M. "Reservoir Development Using Offset VSP Techniques in the Denver-Julesburg Basin," *Journal of Petroleum Technology*)

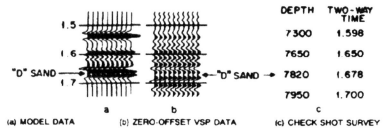

Fig. 6.70 Correlation between model data and zero-offset VSP (Copyright © 1988, Society of Petroleum Engineers, Cramer. P. M. "Reservoir Development Using Offset VSP Techniques in the Denver-Julesburg Basin," *Journal of Petroleum Technology*)

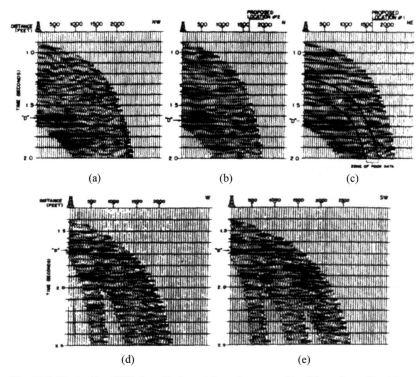

Fig. 6.71 Final offset VSP data displays; (**a**) northwest profile, (**b**) north profile, (**c**) northeast profile, (**d**) west profile, and (**e**) southwest profile (Copyright © 1988, Society of Petroleum Engineers, from Cramer, P.M.: "Reservoir Development Using Offset VSP Techniques in the Denver-Julesburg Basin," *Journal of Petroleum Technology* (February 1988))

Recommendations for Use of Multi-Offset VSP

1. Conduct seismic modeling of the proposed survey before the survey to

   a. verify that the resolution required to solve the problem can be obtained,
   b. assist in designing survey parameters such as source offset and geophone level increment, and
   c. assist the interpreter in understanding the record area.

2. Use a completely cased hole or in the open hole before casing is set, because poorly cemented casing causes serious degradation of the VSP data quality.
3. Design multi-offset VSP surveys to use all possible existing well control to confirm modeled results.
4. Run a near offset VSP with the far offset to establish velocity control to aid in correlating data with the well logs.

VSP is one of the geophysical methods that can be applied to develop a field economically.

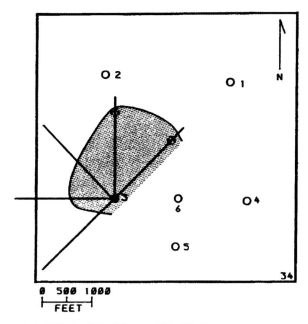

Subsurface Maps

Much of the results of seismic data interpretation are presented as subsurface maps. In the past these were manually constructed but today are generally computer-generated. The techniques and nomenclature are, however, essentially the same. The following is presented to acquaint the student with the types of maps generated and how they are read.

Subsurface Structure Maps

How is a three-dimensional earth to be represented on a two-dimensional surface? Artists do this by the use of perspective. Technical applications, however, require more accuracy than a drawing or painting can supply. The top of Fig. 6.73 shows an artist's sketch of a coastal area where a stream flows through eroded hills into a small bay The bottom of the Figure shows how the same area is represented by a topographic map . The lines with numbers shown on the topographic map are called contours or contour lines. They represent points of equal elevation. The spacing of

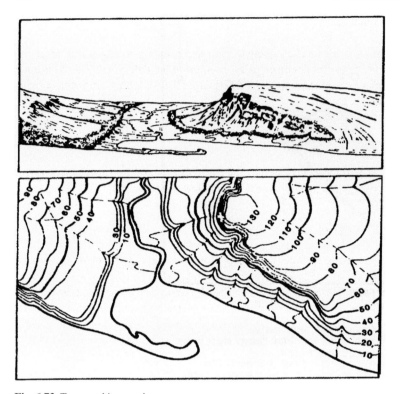

Fig. 6.73 Topographic mapping

the contour lines is a measure of the steepness of the slope; the closer the spacing, the steeper the slope.

A subsurface structural map shows relief on a subsurface horizon with contour lines that represent equal depth below a reference datum or two-way time from the surface. These contour maps reveal the slope of the formation, structural relief of the formation, its dip, and any faulting and folding.

In constructing a subsurface map from seismic data, as mentioned above, a reference datum must first be selected. The datum may be sea level or any other depth above or below sea level. Frequently, another datum above sea level is selected in order to image a shallow marker on the seismic cross-section, which may have a great impact on the interpretation of the zone of interest.

Contouring Techniques

There are three contouring techniques in general use:

- Mechanical spacing.
- Uniform spacing.
- Interpretative contouring.

Mechanical Spacing

Mechanical spacing may be done using a computer-contouring program in which the contour lines are drawn in a mathematical relation to the data point. This technique may be very misleading and some modifications of the computer-generated contour maps are usually required.

Uniform Spacing

Uniform spacing may produce pretty drawings but it is not logical contouring and does not really show the geology of the subsurface correctly.

Interpretive Contouring

This is the technique used for drawing structure contour maps. Knowing the general character and form of geologic structures where the maps are drawn helps to correctly interpret the subsurface, especially where the well control is sparse. Contours should be drawn such that the structure pattern bears out regional trends or tendencies. In this technique, contours are usually drawn parallel to each other.

Figure 6.74 illustrates the three general types of contour drawing.

Mechanical Spacing

Uniform Spacing

Interpretive Contouring

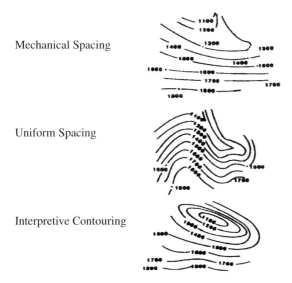

Fig. 6.74 Contouring techniques

Contour Drawing

There are a number of general rules and some basic techniques that are followed in constructing structure maps. These are summarized below.

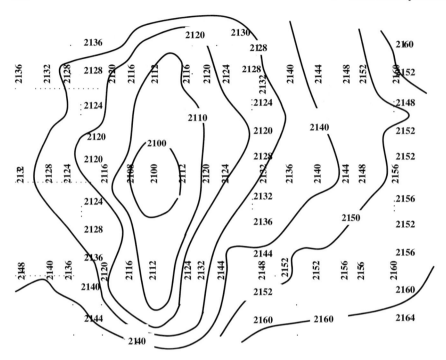

Fig. 6.75 General contouring rules (numbers on map are two-way times in ms)

- Contour lines pass *between* points whose numerical values are higher on one side and lower, on the other, as shown in Fig. 6.75.
- A contour line never crosses over itself or another contour (exceptions – overturned folds and reverse faults).
- The highest or the lowest contour should be repeated where the slope of a structure reverses direction (as in a ridge or valley). See Fig. 6.76.
- Contour lines should not merge with contours of different values or with different contours of the same value. Note that contours sometimes appear to merge when a steeply sloping surface is projected onto a map, the (see Fig. 6.77).
- Contours are usually drawn as closed shapes. However, at the edges of a map there may not be enough data. To close a contour and in the case of faults (see Fig. 6.78) a contour may end abruptly at the fault plane.
- It is helpful in reading a contour map if every fifth contour line is drawn more heavily than the others. Numbers on the contours that are evenly distributed over the map also make reading a map easier.
- The same contour interval should be used over an entire prospect area, whatever contour interval is selected,
- In a stream or a valley, V-shaped contour lines points upstream.
- Dashed lines may be used for contours when control points are lacking.

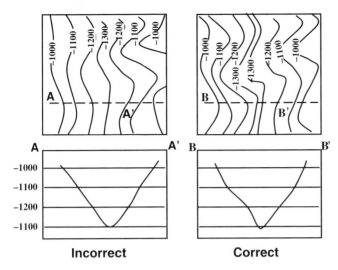

Fig. 6.76 Contouring lows and highs

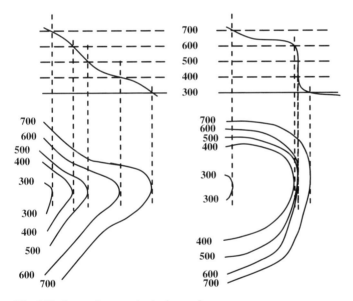

Fig. 6.77 Contouring steeply sloping surfaces

Seismic Structure Maps from Horizontal Slices

Figure 5.75 of Chap. 5 (repeated here as Fig. 6.79) illustrates the use of time slices to produce a seismic structure map. Data were sampled at 4 ms so a time slice or horizontal time section is produced for every 4 ms of two-way time. In the example

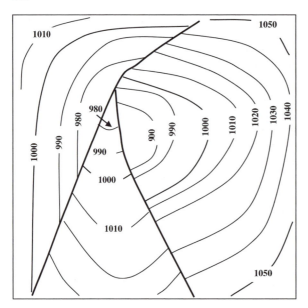

Fig. 6.78 Contours in the presence of faults. Contour interval 10 ms

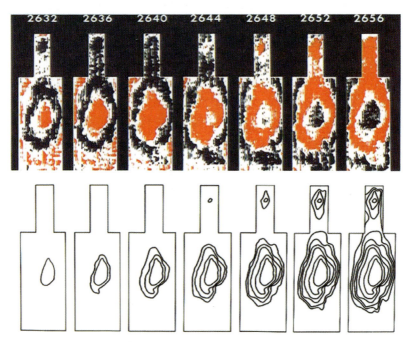

Fig. 6.79 Structure map derived from sequence of time slices 4 ms apart (courtesy of Occidental Exploration and Production Company)

shown, the earliest time for the selected horizon is 2632 ms. The enclosed red area in the center represents amplitudes in the maximum positive amplitude range and the outline of this area is drawn. This is the 2632 ms contour.

The next slice (2636 ms) shows a larger red area near the center. An outline of this is drawn around 2632 ms contour. The process continues with contours drawn at each 4 ms. Since two-way time is being plotted, it can be seen that the 2632 ms contour represents a high. A secondary high occurs at 2644 ms.

If a time-to depth conversion is made, depth slices can be produced. These will be sampled in depth so the contours for this horizon would likely differ somewhat from the time contours.

Isotime and Isopach Maps

I*sopach* and *isochron,* or *isotime,* maps are used to show thickness variations between two subsurface horizons. So, contours, in these maps, are lines connecting points of equal thickness. On isopach maps contours are annotated in feet or meters. On isochron maps contours are annotated in seconds of two-way time differences.

Isopach and isochron maps can be thought of as showing paleostructure; that is, the structure of the lower horizon at the time the upper horizon was deposited. It is a time-interval map that shows the time difference between two markers. Figure 6.80 illustrates this kind of seismic map.

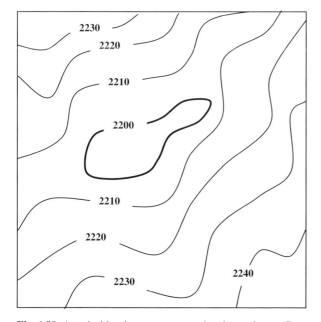

Fig. 6.80 A typical isochron or two-way time interval map. Contour interval 10 ms

Velocity Gradient Maps

Velocities obtained from check shots at wells are accurate, but wells are usually far apart. However, changes in lithology, structure, and other factors can cause velocities to change over even short distances.

Velocities determined from seismic data can be used to supplement velocities from wells but these velocities, even from the best stacking velocity, are not as accurate. Furthermore, accuracy of velocities determined from surface seismic data decrease with depth. If these velocity values are used to convert from time to depth, there may be some problems tying to wells.

It is usually better to map in time rather than in depth, but there will be occasions when, in spite of all the problems, you do need to map in depth. However, engineers, geologists, and managers who want to know the depth. In the development stage of a field, depth maps are more valuable than time maps.

Problems such as velocity pull-up or velocity pull-down cause time structure below the feature that causes the velocity effect to be incorrect. Also, sudden changes in velocities cause the apparent structure on a time map to be quite different from that seen on depth maps. If the velocity to a formation varies, a velocity map can be made to the formation at each well or at points where velocity information is fairly good. Contours can be made to fit consistently (Fig. 6.81).

Mapping velocity variation even with velocities derived from seismic surface data can be informative and useful. Correlations between structure and velocity spatial variation can be investigated to determine their validity.

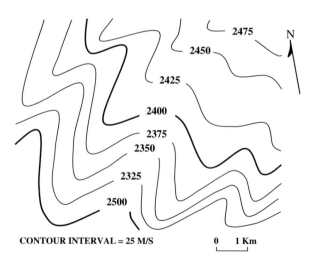

CONTOUR INTERVAL = 25 M/S   0   1 Km

Fig. 6.81 Average velocity map

Interactive Interpretation

Computer hardware and software systems are becoming more and more sophisticated. Current hardware systems are very fast and user friendly. Software systems are also very impressive and can be applied to seismic analysis, reservoir modeling, reservoir simulation, and data management.

Computers continue to get smaller and faster. Desktop systems can be used to perform tasks that once required a mainframe computer – tasks such as monitoring complex reservoir configurations in three dimensions. Most of these programs are easy to use with on-screen guides to help the user through the most complicated applications.

The increased production cost makes understanding reservoir characterization ever more critical. Fortunately, advances in computer graphics software are helping to satisfy that need through the use of 3-D imaging software. Interactive interpretation software allows the user to experiment with a number of possibilities and view the effects on screen almost instantly. This is a great help in solving a structural or stratigraphic problem.

3-D seismic surveys generate huge amounts of data. Processing, sorting, and retrieving these data requires hours of computer. The efficiency of these operations has a large impact on profitability. Many systems are available that can help make exploration data management efficient and productive. Users of these systems can view, select, and request any item from the exploration data base at any point of the operation. Items available include well logs, seismic lines, base maps, etc.

Work stations have proven to be very effective in interpretation of 3-D seismic data. Three-dimensional computer graphics have been integrated with measurement, analysis, interpretation and other reservoir properties to help the interpreter better understand more about the reservoir.

Super computers provide not only extremely fast conventional processing of seismic data but can also perform tasks that would overtax work stations or desk-top computers. For example, super computers generate tomograms that show variations in seismic velocities used for depth conversion and migration. They can generate other tomograms that show variations of petrophysical properties such as porosity, permeability, and fluid content between wells.

Interactive interpretation helps reduce turnaround time in data processing and interpretation. In so doing, it enables oil companies to manage vast amounts of data. This saving of time and effort allows them to meet deadlines and commitments for drilling, lease sale, or monitoring secondary recovery project.

Summary and Discussion

In this chapter, the techniques of modeling, tomography, AVO, and VSP interpretation have been presented and described.

Modeling

Synthetic seismograms are a form of one dimensional modeling and generating a synthetic seismogram is a kind of forward modeling. Properties of rock strata such as velocity and density are used to generate spike traces that are convolved with a reflection wavelet to form a seismic trace. By varying the thickness, replacing, or eliminating some of the geological units used to generate synthetic seismograms an interpreter can see the effect of a change in a geological section and look for similarities in recorded data.

Two-dimensional models assist analysis of geologic interpretations. They can be used to evaluate observed anomalies such as bright spots and dim spots. Two-dimensional models also help solve interpretational problems such as near-surface formation irregularities and velocity pull-ups. Additionally, they can help in designing both field parameters and data processing sequences.

The greatest advantage of modeling may be its application to investigate seismic distortions and their effect on the real subsurface image. Modeling is an excellent tool in educating the explorationist in ways that will help upgrade data interpretation.

Synthetic seismograms and two-dimensional models can be generated on desktop computers. The advances in computer hardware and software allow these very important interpretational tools to be generated and manipulated quickly and easily. Cost of a modeling package varies, depending on the hardware configuration and software sophistication, but is small compared to the value added to seismic interpretation.

Tomography

There are two methods of tomography being used, reflection and transmission. Reflection (travel time) tomography is used to estimate velocities from seismic reflection times. These velocities can be applied to seismic imaging such as depth conversion and pre-stack depth migration.

Two types of transmission tomography are in use: borehole-to-borehole or borehole-to-surface. In the borehole-to-borehole type, a source is placed in one borehole and a receiver in the other. Energy sources can be lowered into one hole and receivers into surrounding wells, allowing velocities to be measured and mapped between wells. These velocities can then be related to the reservoir properties of porosity, permeability, and fluid content. Maps of these properties between wells, can be used to account for vertical and horizontal changes in these parameters. Engineers and geologists can use this information to understand more about the heterogeneity of the reservoir rock and more accurately describe reservoir characteristics.

The combination of tomography and borehole measurements can be the key to success in improving hydrocarbon recovery methods and enhanced oil-recovery projects. This approach requires integration of geophysical, geological and engineering methods. Hence it needs the efforts and effective communications of all disciplines involved.

AVO

AVO analysis, in which changes of amplitude with the angle of incidence are analyzed, is widely used as a hydrocarbon indicator. Common midpoint (CMP) gathers, corrected for NMO, but before stacking are used for the analysis. AVO analysis is two-dimensional compared to stacked traces, which are one-dimensional. AVO techniques can be used to delineate look-alike features, to define depositional environments, delineate reefs, and identify gas sands to name few applications.

A long spread is desirable in the field data acquisition so that the far traces (long distances from the source) can be more readily investigated for changes of the amplitude with offset How long the spread should be depends on target depth, velocity in the area, structure, and maximum frequency to be recorded. Spread length can be determined by field tests or based on the contractor's experience in the area. Seismic modeling could help decide on spread length.

Data used to investigate the variation of amplitude with offset require a special data processing sequence. Only those processes that preserve the relative true amplitude of the seismic traces within the CMP is to be used.

AVO analysis is used to identify the rock lithology and its fluids and/or gas content. The rock properties in a nearby field, velocity information, and stratigraphy should be known in order to perform a reasonable AVO interpretation.

Data processing should be closely monitored, and CMP gathers closely examined for amplitude changes. Thorough investigations should make use of displays such as partial stacks and ratio sections. Analysis and interpretation of amplitude changes should be complemented with modeling that incorporates geological, geophysical, and petrophysical information.

Many AVO processing packages are available. Data can be manipulated to analyze a variation of one variable such as Poisson's ratio, gradient, and others. Cost of producing an AVO analysis is minimal. They may be included in the basic processing sequence.

VSP Interpretation

Vertical seismic profiling (VSP) has proved its value in applications to petroleum exploration and development. The cost of a VSP was substantial a few years ago but now cost substantially less. The turnaround time is a few days and sometimes overnight in case of emergency. The survey is done routinely as any logging tool. In a vertical survey eight to ten levels per hour can be surveyed. In land surveys, perhaps six to eight levels per hour can be taken because it takes more time to inject the energy source. In offshore surveys, it may take more time, four to five levels per hour.

A VSP survey provides the geophysicist with seismic velocity, seismic time to geological depth conversion, and the next seismic marker. It provides the geologist with well prognosis. It will tell the engineer the location of the drilling bit or at what depth he can expect a high-pressure zone. If he can predict the high-pressure zone ahead of time, he can take action to head off problems. With minimal rig idle time, the survey is definitely more economical than a blowout.

The VSP will play an important role in borehole geophysics, reservoir characteristics, and transmission tomography.

Subsurface Maps

While modern computer graphics used in processing and interpretation of seismic data have all but eliminated manual mapping. The steps in generating these continue the same. Thus, today's geoscientist needs to understand what the displays are and how to read them.

Interactive interpretation software run on immensely powerful and fast work stations allow the user to view 3-D seismic data volumes in many ways, including true 3-D visualization. However, old-fashioned structure maps, isopach maps and velocity gradient maps still have a place in the total package. Knowing about them and how to read them may be valuable information for the reader.

Bibliography

Aki, K., and P. G. Richards. *Quantitative Seismology*. Cranbury, NJ: W. H. Freeman and Co., (1980).

Anstey, N. A. "Attacking the Problems of the Synthetic Seismogram." *Geophysical Prospecting 8* (1960):242–260.

Anstey, N.*A Seismic Interpretation: The Physical Aspects,* Boston: II3DRC, (1977).

Anstey, N.*A Seismic Exploration for Sandstone Reservoirs,* Boston: II3RDC, (1978).

Arya, V. K. and H. D. Holden. *A Geophysical Application: Deconvolution of Seismic Data.* N. Hollywood, CA: Digital Signal Processing. Western Periodicals, (1979):324–338.

Badws, M. M. "The Reflection Seismogram m a layered Faith" *Bulletin of the American Association of Petroleum Geologists 67* (1983):416–417.

Baranov, V. "Film Synthetique Avec Reflexions Multiples-Theorie Et Calcul Practique." *Geophysical Prospecting 8* (1960):315–325.

Barton, D. C. "The Seismic Method of Mapping Geologic Structure." *Geophysical Prospecting American Institute of Mining and Matallurgical Engineering* (1929): 572–624.

Beydoun, W. B., J. Delvaux, M. Mendes, G. Noual,, and A Tarantola. "Practical aspects of an elastic migration/inversion of crosshole data for reservoir characterization: a Paris Basin example." *Geophysics 54* (1989):1587–1595.

Bilgeri, D. and E. B. Ademeno. "Predicting Abnormally Pressured Sedimentary Rocks." *Geophysics 30* (1982):608–621.

Biot, M. A. "Propagation of Elastic Waves in a Cylindrical Bore Containing a Fluid." *Journal of Applied Physics 23* (1952):997–1005.

Bishop, T. N., Bube, K. P., Cutter, R. T., Langen, R. T., Love, P. L., Renick, J. R., Shuey, R. T., Spindler, D. A., and Wyll, H. W. "TomographicDetermination of Velocity and Depth in Laterally Varying Media", *Geophysics 50* (1985):903–923.

Bois, P., M. Laporte, M. Laverne, and G. Thomas. "Well-to-Well Seismic Measurements." G*egphysics 37* (1972):471–480.

Bording, R. P., A. Gersztenkorn, L. Lines, J. Scales, and S. Treitel. "Applications of Seismic Travel-time Tomography." *Geophysics Journal of the Royal Astronomical Society 90* (1987):285–303.

Bregman, N. D., R. C. Bailey, and C. H. Chapman. "Cross-Hole Seismic Tomography." *Geophysics 54* (1989):200–215.

Brewer, H. L. and J. Holtzscherer. "Results of Subsurface Investigations Using Seismic Detectors and Deep Bore Holes." *Geophysics Pro-T 6* (1958):81–100.

Butler, D. K. and J. R Curro Jr. "Crosshole Seismic Testing-Procedures and Pitfalls." Geophysics *46* (1981): 23–29.

Castagna, J. P. "Petrophysical Imaging Using AVO." *TLE12* (1993):172–178.

Castagna, J. P., M. L. Batzle, and R. L. Eastwood. "Relationships Between Compressional Wave and Shear-Wave Velocities in Clastic Silicate Rocks." *Geophysics50* (1985):571–581.

Chun, J., D. G Stone, and C. A. Jacewitz. *Extrapolation and Interpolation of VSP Data.* Tulsa, OK: Seismograph Service Companies Reprint, (1982).

Coffeen, J. A. *Interpreting Seismic Data.* Tulsa, OK: PennWell Publishing Company, (1984).

Coffeen, J. A. *Seismic On Screen.* Tulsa, OK: PennWell Publishing Company, (1990).

Collins, F. and C. C. Lee. "Seismic Wave Attenuation Characteristics from Pulse Experiments." *Geophysics 21* (1950):16–40.

Delaplanhce, J., R. F. Hagemann and P. G. C. Bollard. "An Example of the Use of Synthetic Seismograms." *Geophysics 28* (1963):842–854.

Dennison, A T. "An Introduction to Synthetic Seismogram Techniques." *Geophysical Prospecting 8* (1960):231–241.

DeVoogd, N. and H. Den Rooijen. "Thin layer response and spectral bandwidth." *Geophysics 48* (1983):12–18.

Dines, K. A. and R. J. Lyfle, "Computerized Geophysical Tomography." *Proc. IFFF, 67* (1979):1065–1073.

Domenico, S. N. "Effect of Brine-Gas Mixture on Velocity in an Unconsolidated Sand Reservoir." *Geophysics 41* (1976):882–894.

Durschner, H. "Synthetic Seismograms from Continuous Velocity Logs." *Geophysical Prospecting 6* (1958):272–284.

Dyer, B. C. and M. H. Worthington. "Seismic Reflection Tomography: a Case Study." *First Break 6* (1988):354–366.

Faust, L. Y. "A Velocity Function Including Lithologic Variation." *Geophysics 18* (1953):271–288.

Futterman, W. I. "Dispersive Body Waves." Journal of *Geophysical Research 67* (1962):5279–5291.

Gaizer, J. E. and J. P. DiSiena. "VSP Fundamentals That Improve CDP Data Interpretation." Paper S 12.2, 52nd Annual International Meeting of SEG, Technical Program Abstracts, (1982a):154–156.

Gardner, G. H. F., L. W. Gardner and A. R. Gregory. "Formation Velocity and Density the Diagnostic Basics for Stratigraphic Traps." *Geophysics 39* (1974):770–780.

Gassaway, G. S. and H. J. Richgels. "SAMPLE, Seismic Amplitude Measurement for Primary Lithology Estimation." *53rd SEGMtg.,* Las Vegas, Expanded Abstracts (1983):610–613.

Gelfand, V., P. Ng, and K. Lamer. "Seismic Lithologic Modeling of Amplitude-Versus-Offset Data." *56th SEG Mtg.,* Houston, Expanded Abstracts (1986):332–334.

Gerritsma, P. H. A. "Time to Depth Conversion in the Presence of Structure." *Geophysics 42* (1977):760–772.

Goupillaud, P. L. "An Approach to Inverse Filtering of Near-Surface Layer Effects From Seismic Records." *Geophysics 26* (1961):754–760.

Han, D., A. Nur, and D. Morgan. "Effects of Porosity and Clay Content on Wave Velocities in Sandstones." *Geophysics 51* (1986):2093–2107.

Hill, N. R., and I. Lerche. "Acoustic Reflections From Undulating Surfaces." *Geophysics 51* (1986):2160–2161

Hilterman, F. J. "Three-Dimensional Seismic Modeling." *Geophysics 35* (1970):1020–1037.

Hilterman, F. J. "Amplitudes of Seismic Waves-A Quick Look." *Geophysics 40* (1975):745–762.

Hilterman, F. "Is AVO the Seismic Signature of Lithology? A Case History of Ship Shoal South Addition." *TLE3* (1990):15–22.

Hilterman, F. "AVO: Seismic Lithology." *SEG, Course Notes* (1992).

Hindlet, F. "Thin Layer Analysis Using Offset/Amplitude Data." *56th* SEGMtg., Houston, Expanded Abstracts (1986):332–334.

Ivansson, S. "Seismic Borehole Tomography-Theory and Computational Methods." *Proc. IEEE 71* (1986):328–338.

Ivansson, S. "A Study of Methods for Tomographic Velocity Estimation in the Presence of Low Velocity Zones." *Geophysics 56* (1985):969–988.

Lines, L. R. and E. D. La Fehr, "Tomographic Modeling of a Cross-Borehole Data Set", *Geophysics 54* (1989): 1249–1257.

Lines, L. R "Applications of Tomography to Borehole and Reflection Seismology." TLE *10* (1991):11–17.

Lytle, R. J. and M. R. Portnoff. "Detecting High-Contrast Seismic Anomalies Using Cross Borehole Probing." *IFRF, Transactions Geasci: Remote Sensing 22* (1984):93–98.

Marcides, C. G., E. R. Kanasewich, and S. Bharatha. "MMultiborehole Seismic Imaging in Steam Injection Heavy Oil Recovery Projects." *Geophysics 53* (1988):65–75.

McMechan, G. A. "Seismic Tomography in Boreholes." *Geophysical Journal of the Royal Astronomical Society 74* (1983):601–612.

Meissner, R. and M. A. Hegazy. "The Ratio of the PP- to SS-Reflection Coefficient as a Possible Future Method to Estimate Oil and Gas Reservoirs." *Geophysical Prospecting 29* (1981):533–540.

Muskat, M. and M. W. Meres. "Reflection and Transmission Coefficients for Plane Waves in Elastic Media." *Geophysics 5* (1940):149–155.

Mufti, I. R. "Numerical Experiments with a Salt Dome," *Geophysics 54* (1989):1043–1045.

Mufti, I. R., "Large-Scale Three-Dimensional Seismic Models and Their Interpretive Significance," *Geophysics 55* (1990):1166–1182.

Mufti, I. R., "Pitfalls in Crosshole Seismic Interpretation as a Result of 3-D Effects," *Geophysics* (1995):821–833.

Mustafayev, K. A. "Increased Absorption of Seismic Waves in Oil and Gas Saturated Deposits." *Prikladnaya Geofizika 47* (1967):42.

Myron, J. R., L. R. Lines, and R. P. Bording, "Computers in Seismic Tomography." *Computers in Physics 52* (1987):26–31.

Narvey, P. J. "Porosity Identification Using AVO in Jurassic Carbonate, Offshore Nova Scotia." *TLE V1* (1993):180–184.

Nur, A. "Four-Dimensional Seismology and (True) Direct Detection of Hydrocarbons: the Petrophysical Basis." *The Leading Edge 8* (1989):30–36.

Ostrander, W. J. "Plane Wave Reflection Coefficients for Gas Sands at Non-Normal Angles of Incidence." *Geophysics 49* (1984):1637–1648.

Peterson, R. A., W. R. Fillipone, and F. B. Coker. "The Synthesis of Seismograms From Well Log Data." *Geophysics 20* (1955):516–538.

Pettijohn, F, J. *Sedimentary Rocks,* New York: Harper & Row, (1948).

Pirson, S. J. *Geologic Well log Analysis,* Houston, Texas: GPC, (1970).

Pucket, M. "Offset VSP: A Tool for Development Drilling." *TLEIQ* (1991):18–24.

Schoenberger, M. and F. K. Levin. "Reflected and Transmitted Filter Functions for Simple Subsurface Geometries." *Geophysics M* (1976):1305–1317.

Sengbush, R L., P. L. Laurence, and F. J. McDonal. "Interpretation of Synthetic Seismograms." *Geophysics 26* (1961):138–157.

Sheriff, R. E. "Encyclopedic Dictionary of Exploration Geophysics." Tulsa, OK: *Society of Exploration Geophysicists* (1973).

Sheriff, R. E. "Factors; Affecting Amplitudes-A Review of Physical Principles, in Lithology and Direct Detection of Hydrocarbons Using Geophysical Methods." *Geophysical Prospecting 25* (1973):125–138.

Sheriff, R E. *A First Course In Geophysical Exploration and Interpretation.* Boston: IHDRC, (1978).

Shuey, R. T. "A Simplification of Zoeppritz Equations." *Geophysics 50* (1985):609–614.

Stewart, R. R., R. M. Turpenmg and M. N. Toksoz. "Study of a Subsurface Fracture Zone by Vertical Seismic Profiling." *Geophysical Research Letters 9* (1981):1132–1135.

Stone, D. G. "VSP-The Missing Link" Paper presented at the VSP Short Course Sponsored by the Southeastern Geophysical Society in New Orleans, (1981).

Stone, D. G. "Prediction of Depth and Velocity on VSP." Paper presented at the 52nd Ann Mtg. of SEG, Dallas, Toms, (1982).

Stone, D. G. "Predicting Pore Pressure and Porosity From VSP Data." Paper presented at the 53rd Ann. Mtg. of SEG, Las Vegas, Nevada, (1983).

Treitel, S. "Seismic Wave-Propagation in Layered Media in Terms of Communication Theory." *Geophysics 31* (1966):17–32.

Treitel, S. "Quo vadit inversion?" *The Leading Edge 8* (1989):38–42.

Trorey, A. W. "Theoretical Seismograms With Frequency and Depth Dependent Absorption." *Geophysics 27* (1962):766–785.

Wharton, J. B., Jr. "Isopachous Maps of Sand Reservoirs." *Bulletin of the AAPG 32* (7) (1948):1331–1339.

Wiggens, R A., K. L. Lamer, and R. D. Wisecup. "Residual Statics Analysis as a General Linear Inverse Problem." *Geophysics 41* (1976):922–938.

Wong, J., N. Bregnan, G. West, and P. Hurely. "Cross-Hole Seismic Scanning and Tomography." *The Leading Edge 6* (1987):36–41.

Worthington, M. "An Introduction to Geophysical Tomography." *First Break 2* (1984):20-27.

Wright, J. "Reflection Coefficients at Pore-Fluid Contacts as a Function of Offset." *Geophysics 51* (1986):1858–1860.

Wuenschel, P. C. "Seismogram Synthesis Including Multiples and Transmission Coefficients." *Geophysics 25* (1960):106–219.

Wyatt, K. D. "Synthetic Vertical Seismic Profile." *Geophysics 46* (1981a):880–991.

Young, G. B. and L. W. Braile. "A Computer Program for the Application of Zoeppritz's Amplitude Equations and Knott's Energy Equations." *Bulletin of the Seismological Society of America 66* (6) (1976):1881–1885.

Yu, G. "Offset-Amplitude Variation and Controlled-Amplitude Processing." *Geophysics 50* (1985):2697–2708.

Zhu, X, P. O. Sixta, and B. G. Atstman. "Tomostatics: Turning Ray Tomography + Static Corrections." *TLE II* (1992):15–23.

Chapter 7
4-D (Time Lapse 3-D) Seismic Surveys

Introduction

The increase in demand for hydrocarbons all over the world, coupled with political unrest in some oil producing countries, makes oil price escalate to new highs. As a result, offshore exploration in deep water has become more attractive than ever before. However, there is still a time gap between the development of the fields and production. Note that it sometimes is not economical.

There are already producing fields in which primary production has been completed. The efficiency of production ranges from 30 to 50% of the hydrocarbons in place. In reservoir evaluation and prediction of future production, the formation is considered to be homogeneous, i.e., it is assumed that all particles of the reservoir have the same properties. No account is taken for the lateral and vertical variations of the petrophysical properties of the formation. In many cases, knowledge of the subsurface structure is not detailed enough because of the limitations inherent in 2-D seismic data.

These pitfalls create demand for better resolution from the seismic data and better understanding of the reservoir properties. The industry responded by developing digital recording system that can record more than ten thousand channels. 3-D seismic surveys were introduced with much closer spacing of the geophones. Seismic boats are now well equipped with navigation instruments and GPS to increase the accuracy of the offshore seismic source and hydrophone locations. The ships are now able to tow more than ten streamers thereby increasing subsurface coverage and seismic resolution.

However, to understand more about the reservoir, it is necessary to go down to the reservoir and record seismic information such as changes in velocities and relate these changes to the other petrophysical properties of the reservoir. The feasibility of secondary or tertiary recovery projects in currently producing fields can be explored via 3-D surveys and borehole measurements that use one well as a source and another well or wells for receivers. Of course, cost of these projects against potential production increase must be considered.

As mentioned in Chap. 6, the transmission tomography method can be applied to generate tomograms that show the variation of porosity, permeability, and water content to name a few reservoir properties. Generating a tomogram takes some

M.R. Gadallah, R. Fisher, *Exploration Geophysics*,
DOI 10.1007/978-3-540-85160-8_7, © Springer-Verlag Berlin Heidelberg 2009

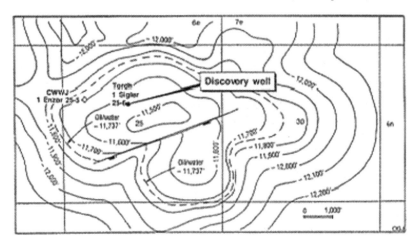

Fig. 7.1 Manually drawn structure map based on data from twenty 2-D seismic lines

computer time but it is becoming more affordable. With the clearer picture of sub-surface structure, provided by 3-D seismic surveys and borehole seismic transmis-sion between wells, reservoir properties are better understood and can be used to estimate the hydrocarbon in place. The 3-D survey reveals finer details that the 2-D survey never reveals because of its limited resolution. These details can be small faults, lithology changes, permeability barriers, etc., that alter the geometry of the reservoir. The project involves all the specialists from disciplines - geophysicists, ge-ologists and reservoir engineers. Each of these professionals contributes their share of knowledge to the other member of the team.

This new techniques can improve production efficiency by more than 10%, dras-tically increasing the amount of recoverable hydrocarbons.

To illustrate the approach, Figs. 7.1 and 7.2 show how much more detail can be seen in the structure map interpreted from the 3-D seismic survey than from 2-D.

Enhanced Oil Recovery (EOR)

A number of techniques allow mapping of EOR processes. One of the more famil-iar techniques uses 3-D seismic, borehole seismic, and micro-seismic. Defining the EOR front depends on the change in density of the reservoir rock as a result of the recovery process. Commonly used techniques that can induce substantial density changes in the rock include steam flooding and in-situ combustion. The particular seismic method used depends on what is to be accomplished. For example, a com-plete description of the developing front can be obtained by using 3-D seismic map-ping, while limited knowledge can be obtained by using conventional 2-D seismic along the line.

Several considerations are involved in field application of 3-D seismic map-ping in enhanced oil-recovery projects using steam injection. Working with this

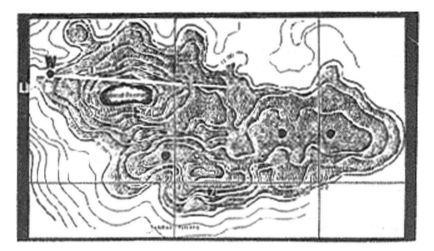

Fig. 7.2 Structure map from 3-D data

technique is expensive in terms of both fieldwork and data reduction. The process may have to be shut in during tests. A large number of stations and high-frequency sources are required to obtain the necessary resolution. The large amount of data and special processing require longer time. The surveys do, however, provide a clearer and more accurate solution, since time-slice maps and sections can be obtained in any direction through the 3-D data sets.

This technique can be used to monitor the changes and map the swept zone. Information is obtained that an engineer can use to optimize production through better control, foresee developing problems, and be able to take the proper corrective action in a timely fashion.

Figure 7.3 illustrates the 3-D seismic survey as it was used to map the steam flood project at Street Ranch Test. The recording pattern consisted of four seismic survey lines, each of which ran through the center injection well of an inverted five-spot pattern.

In order to facilitate the seismic survey, the pilot was shut in for three days. The formation was 460 m (1,500 ft) in depth. The pay zone was 16 m (50 ft) thick. However, steam was injected only in the upper 8 m (25 ft) of the zone. A VSP survey was conducted to help determine the reflection characteristics of the target formation. A high-frequency energy source was used to obtain high resolution of the EOR swept front.

In Fig. 7.3(b) there are four slices taken at lines 3, 1, 4 and 2 around the target formation. The changes in reflection seismic data were due to changes in reservoir impedance caused by an increase in gas saturation. The seismic sections show definite changes in wavelet shape over the center of the injection well and extend partially to the production wells. In the zone where steam was injected, a second peak developed on top of the peak of the pay.

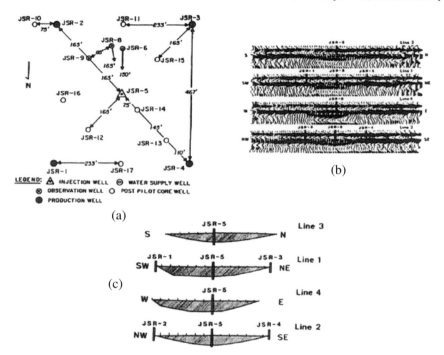

(a)

(b)

(c)

Fig. 7.3 3-D seismic mapping of steam flood – street ranch pilot test. (Courtesy of society of petroleum engineers)

The most useful approach is to carry out a series of surveys at time intervals that allow the development of the EOR process to be established. This requires surveys at three-month to six-month intervals. One of the problems of the resolution can be amplified by the heterogeneous nature of the reservoir geology, especially with the surface techniques of conventional 3-D seismic survey. Caution must be exercised in establishing the geometry of the observation survey layout. It is possible to miss critical developments by improper placement of source and receiver. In addition, if the response measured, such as wave characteristics, is not related in a linear manner to the EOR process, it is possible to misinterpret the data. This makes it advisable to perform laboratory measurements that will aid in the interpretation of the field measurements.

Bibliography

Britton M. W. et al. "The Street Ranch Pilot test of Fracture-Assisted Steamflood Technology," *Journal of Petroleum Technology* March, (1983).

Chapter 8
Future Trends

Currently, the major concerns in the oil industry are in recovering more hydrocarbons from presently producing fields and improving the success ratio in exploration, development and exploitation. The reservoir team is the new trend in oil company organization and is the key to success in their efforts to enhance their reserves.

It has become clear that depth imaging of the subsurface must be done to get better resolution of geometry of the geological structure, fault orientation, as will as structure and fine details of the stratigraphy. A reliable velocity model is required to convert seismic data from the time domain into the depth domain. Such models must represent not only the vertical variation of velocity with depth but also lateral variation caused by heterogeneity of the subsurface rock layers.

Using three-component seismic data acquisition both P-wave, SV-wave (vertical component of shear wave) velocities can be determined. This allows various elastic moduli to be determined. The 4C systems (four components – the previous three plus hydrophones for OBC measurement), allow many corrections to seismic data to be applied that improve the signal to noise ratio and enhance the seismic resolution.

Borehole measurements, by lowering seismic source in a well and receivers in other wells, help to derive velocity tomograms that are used to get other tomograms that represent the spatial variation of petrophysical rock properties such as porosity, permeability, clay content, and water saturation to name few. This information will increase our understanding of the reservoir characterization and improve the estimation of the hydrocarbons in place.

Advances in 3-D seismic data acquisition and computer hardware for data processing make it more affordable to conduct the 3-D seismic surveys, even for exploration efforts to solve the complications of the subsurface geological settings. Lengthy turnaround time for data processing is not an issue, as the giant computing ability and more sophisticated, faster software applications make the acquisition and data processing economically feasible.

Tomography, as was discussed in Chap. 6, helps in deriving velocity models for pre-stack depth migration and depth conversion. From the elastic moduli, corrections for change of seismic velocity with direction of propagation (anisotropy) can be determined and applied in selective areas. The measurements of VSP and long offset VSP can be used to correct the seismic data to the well depth and better image the subsurface geology that will help in locating the next drilling location.

M.R. Gadallah, R. Fisher, *Exploration Geophysics*,
DOI 10.1007/978-3-540-85160-8_8, © Springer-Verlag Berlin Heidelberg 2009

4-D surveys might contribute in increasing the recoverable hydrocarbons as better understanding of the reservoir performance can be observed through the behavior of the swept front.

Pooling leases on the same producing reservoir is not as easy task for the domestic oil industry. The merger of oil companies in the past decade will promote the use of 3-D time-lapse seismic surveys to monitor secondary and tertiary recoveries projects.

Deep offshore-drilling, is another contributor to the increase of reserves. Current crude oil prices, at the time this text was written ($50.00–$60.00/bbl), make deep offshore drilling in offshore Angola, South Africa, and the Gulf of Mexico economically feasible The price of crude oil is likely to remain at or above current levels for quite some time.

The oil industry is changing very rapidly in all aspect of exploration, development, and enhanced oil recovery to improve the production efficiency in the current producing oil fields. It is the opinion of the authors that the future of geologists, geophysicists, land people and engineers is promising and bright. They will face new challenges in an ever-changing industry. With the ups and downs in the industry, many experienced professionals left the industry either for manpower reduction or retirement, resulting in an experience gap among the current employees, intensive training across disciplines is needed to close the gap in communication among the reservoir team members.

We are very optimistic that fossil hydrocarbons will be in demand as a source of energy for many decades to come.

Appendix A

To understand F-K migration, it is important to understand migration in the depth domain. Chun and Jerkewitz (1981) explained this method in a very elegant and clear manner.

Depth Domain Migration

Figure A.1a illustrates the vertical earth model ($\theta_a = 90°$).

Consider a seismic source A with a signal recorded in the same point; then the only energy that can be recorded at A is the horizontal path in the ray theory approach. Any non-horizontal traveling wave will be reflected downward and will not return to A. Mapping the distance of the horizontally traveling path in the (X, Z) plane in the Z direction is shown in Fig. A.1b.

Since AO = OC, the dip angle reflector in the record section is equal to 45°. Thus, for a 90° reflector the reflection takes place only at a point on the surface, and the recorded reflections are mapped along a 45° line on the depth plane as A moves along the surface.

Next, consider a dipping earth model as in Fig. A.2a. Assuming that the source and receiver are at point A, the wave from A will be reflected at C' and will be recorded at A. The travel distance is thus AC' = AC. When the earth model in (a) is superimposed on (b) we can see that:

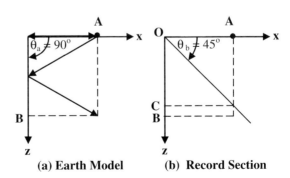

Fig. A.1 90 degree reflector model

(a) Earth Model　　　　**(b) Record Section**

Fig. A.2 Dipping reflector model

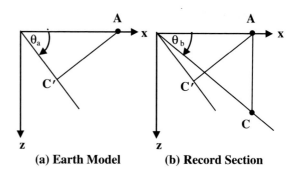

(a) Earth Model   **(b) Record Section**

$$sin\theta_a = AC'/AO = AC/AO = \tan\theta_b$$

This equation describes the relationship between migrated angle (6_a) and recorded angle (d_f). Since point C maps to C' under migration, this process moves data up-dip. b. Diffraction Concept

The concept of diffraction is required to understand migration properly. Diffractions are normally associated with discontinuities, and reflections may be considered as a superposition of diffractions.

So the general process of mapping a reflector or diffractor to the earth model will be described as a diffraction process.

Migration proceeds from a record to the earth model. Diffraction proceeds in the opposite direction, from earth model to record section.

Appendix B

Design of Maximum Offset (Horizontal Reflector Case)

Figure B.1 and Table B.1 explain the calculations of maximum offset for a non-dipping target horizon. One can see that the maximum offset is a function of the depth and angle of incidence. Note that the maximum offset is from source to reflector to receiver point, and it is indicated by $2X$.

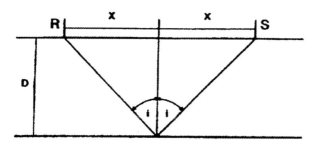

$$\tan(i) = X/D \qquad\qquad i = \text{angle of incidence}$$
$$X = D\tan(i) \qquad\qquad X = \text{half offset distance}$$
$$X_{max} = 2D\tan(i) \qquad\qquad D = \text{depth of target}$$

Fig. B.1 Maximum offset – horizontal reflector

Table B.1 Maximum offset calculations for a horizontal reflector. $X_{max} = 2D\tan(i)$

i°	X_{max}
25	0.90D
30	1.15D
35	1.40D
40	1.68D
45	2.00D

Design of Maximum Offset (Dipping Reflector Case)

Figures B.2 and B.3 show how to calculate the maximum offset of a dipping event.

The offset is a function of depth in unit distance, angle of incidence, and dip angle.

Note: Fig. B.2 is to calculate X, which is the offset distance from source S to the reflection point.

Figure B.3 is to calculate Y, which is the distance from the reflecting point to the receiver. The sum of X and Y *is* the maximum required offset for the case of a dipping target.

To compute the shot line spacing (see Fig. B.4), one must first compute shot interval.

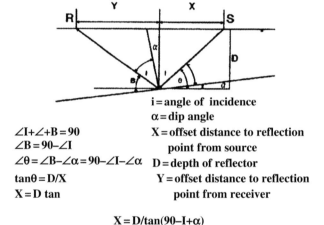

i = angle of incidence
α = dip angle

$\angle I + \angle + B = 90$ X = offset distance to reflection
$\angle B = 90 - \angle I$ point from source
$\angle\theta = \angle B - \angle\alpha = 90 - \angle I - \angle\alpha$ D = depth of reflector
$\tan\theta = D/X$ Y = offset distance to reflection
$X = D \tan$ point from receiver

$$X = D/\tan(90 - I + \alpha)$$

Fig. B.2 Maximum offset – dipping reflector

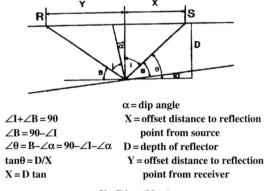

α = dip angle
$\angle I + \angle B = 90$ X = offset distance to reflection
$\angle B = 90 - \angle I$ point from source
$\angle\theta = B - \angle\alpha = 90 - \angle I - \angle\alpha$ D = depth of reflector
$\tan\theta = D/X$ Y = offset distance to reflection
$X = D \tan$ point from receiver

Fig. B.3 Maximum offset –
dipping reflector (2)

$$Y = D/\tan(90 - \alpha)$$

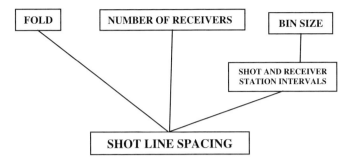

Fig. B.4 Parameters needed to compute line spacing

$$\text{Shot line spacing} = \frac{\text{number of receivers per receiver line times G.I.}}{\text{number of fold in the direction of shot line}}$$

Example: Suppose that the required fold in the receiver line direction is four, and the number of fold in the shot line is six. For maximum fold, $4 \times 6 = 24$ fold. Number of receivers per receiver line is 60 and group interval is 55 ft.

The shot line spacing $= 60 \times 55/6 = 550$ ft. Need to space the shooting lines at an increment of 550 ft.

Appendix C

Answers to Odd-Numbered Exercises

Chapter 3

1. Name and describe three types of seismic waves described in this chapter.
   Answer: Longitudinal or P wave, shear or S wave, Rayleigh wave, and Love wave.
3. Two layers are separated by an interface. The upper layer has a velocity of 2.5 km/s and the lower layer has a velocity of 5.0 km/s. If a ray travels downward through the top layer at an angle of incidence of $20°$, then at what angle will the ray travel in the lower layer? What is the critical angle?

Answer:
$$\sin\phi = \frac{V_L}{V_U}\sin\theta = \frac{5.0}{2.5}\sin 20° = 2 \times 0.34202 = 0.68404$$
$$\phi = 43.1608°$$

$$\text{Critical angle} = \sin^{-1}\left(\frac{V_U}{V_L}\right) = \sin^{-1}\left(\frac{2.5}{5.0}\right) = \sin^{-1}(.5) = 30°$$

5. A horizontal layer has a velocity of 2500 m/s and a thickness of 1500 m. If a detector is placed 1200 m from the source, what is the reflection time?
   Answer: $T = \frac{\sqrt{x^2+4Z^2}}{V} = \frac{\sqrt{1200^2+4\times1500^2}}{2500} = 1.292$ s

Chapter 4

1. How does depth of the charge affect recording of seismic data?
   Answer: Charges above the consolidated rock layer (on or in the weathering or low velocity layer) tend to produce more noise and attenuation.
3. What is the "bubble effect" produced by single airguns? What is done to minimize it?
   Answer: The bubble effect is caused by successive expansions and contractions of the compressed air exhausted from a single airgun. The result is a very long

source wavelet. It is minimized by combining many airguns of different sizes in a geometric arrangement called an airgun array.

5. Identify the natural frequency of the geophones from the amplitude responses shown below.

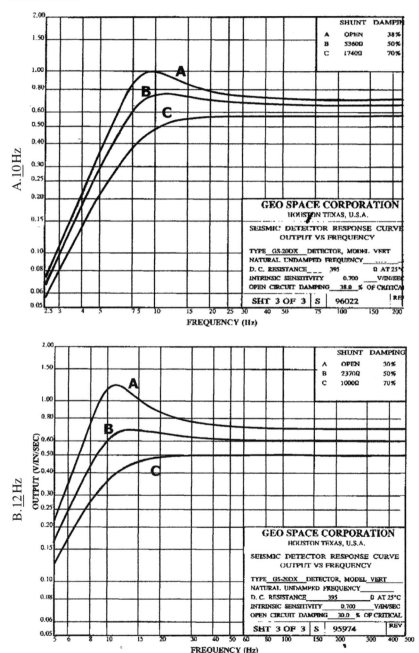

7. Show how decimal 583 is subtracted from decimal1947 in computers.
   Answer: $1047_{10} = 10000010111_2$ $583_{10} = 01001000111_2$

$$1\text{'s complement of } 583_{10} = 10110111000$$

$$+1$$

$$2\text{'s complement of } 583_{10} = 10110111001$$

$1047_{10} = 10000010111$
$-583_{10} = +10110111001$
$\phantom{-583_{10}} = 464_{10} = 00111010000_2$

9. A 192 group spread is to be shot at 24-fold. How many groups must be moved forward between shots to achieve this?
   Answer Must move 1/48 of spread to get 24-fold. So, 192/48 = 4 groups to move between shots.

Chapter 5

1. The maximum frequencies for the data sets A, B, and C are:

   A. 75 Hz
   B. 61 Hz
   C. 129 Hz

   What sample intervals can be used for each data set?
   Answer At 1 ms sampling $f_N = 500$ Hz, at 2 ms $f_N = 250$ Hz, at 4 ms $f_N = 125$ Hz, , at 8 ms $f_N = 62.25$ Hz,
   Therefore, for set A sample period can be 8 ms or less
   for set B sample period can be 4 ms or less
   for set C sample period can be 2 ms or less

3. Find the convolution of A. (4, –2, 1, 3) and B. (–1, 0, 1).
   Answer Let C be the convolution of A and B. Substituting in matrix for convolution (below), cross-multiplying and summing along arrows gives C = –4, 2, 3, –5, 1, 3

5. What is the difference between a front-end mute and a surgical mute? Why are they applied?
   Answer A front-end mute is applied at the start of a record while a surgical mute is applied somewhere in the body of a record. The front-end mute is

applied to eliminate the first breaks and high amplitude events following them. These events provide no subsurface information. Surgical mutes are applied to eliminate noise trains that interfere with signal.

7. From the listed offsets and reflection times given below, do an X^2–T^2 plot. Fit a straight line to the plotted points and estimate velocity, V, and zero-offset time, T_0.

Answer: From the plot below and line fit to points, $T_0^2 = 5.05$ s^2 and $T_0 = 2.247$ s

$$V^2 = \frac{32500000}{8 - 5.05} = 11016950 \text{ m}^2/\text{s}^2$$

$$V = 3319 \text{ m/s}$$

Offset, X (m)	Time, T (s)	X^2	T^2
100	2.251	10000	5.067001
500	2.255	250000	5.085025
900	2.265	810000	5.130225
1300	2.287	1690000	5.230369
1700	2.300	2890000	5.290000
2100	2.338	4410000	5.466244
2500	2.380	6250000	5.664400
2900	2.410	8410000	5.808100
3300	2.460	10890000	6.051600
3700	2.500	13690000	6.250000
4100	2.575	16810000	6.630625
4500	2.620	20250000	6.864400
4900	2.700	24010000	7.290000
5300	2.760	28090000	7.617600
5700	2.830	32490000	8.008900

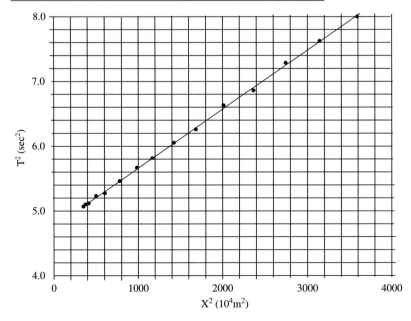

A least-square error solution gives V=3314 m/s and $T_0 = 2.249$ s
9. How does whitening or spiking decon differ from gapped decon?
   Answer: Whitening decon attempts to make the output amplitude spectrum as flat as possible over the entire bandwidth. In so doing it designs inverse filters for the source wavelet, ghosting operators, multiple operators, and instrument response. The whitening process may increase noise outside the signal bandwidth and can introduce false signals. Gapped or prediction error decon affects the source wavelet less or not at all, depending on gap length, but does design inverse filters for the ghosting operators, multiple operators, and instrument response. Whitening decon is sometimes described as a prediction error filter with a zero prediction length.
11. Do band pass filters get rid of all the noise on a seismic record or section? Why or why not?
   Answer: No, band pass filters do not get rid of all the noise on a seismic record or section. They have no effect on noise within the signal frequency band. Outside the signal frequency band the attenuation rate increases so that the farther noise frequencies are from the filter pass band the more they are attenuated. It is impossible to design a filter that zeroes all noise components outside the filter pass band.
13. Why run multiple iterations of velocity and residual statics analysis?
   Answer: The first iteration of velocity and residual statics analysis provides neither exact residual statics nor residual NMO. Successive iterations of velocity and residual statics analysis reduce residual statics and residual NMO to provide optimum velocities and optimum corrections for the near surface.
15. Does post-stack migration cause reflection dip to increase or decrease compared to the stack?
   Answer: Post-stack migration increases the dip of dipping events on seismic sections.

Glossary

Abnormally high-pressure zone   High formation pressure is a pressure is higher than the expected hydrostatic pressure at a certain the formation depth. The normal static pressure gradient is 0.5 psi/ft. Reservoirs that have abnormal pressure usually show low seismic velocity.

Abnormal pressure   Formation fluid pressure that higher than the normal hydrostatic pressure, which is the pressure produced by a column of fluid extending to the surface.

Absorption   It is a process that converts seismic energy into heat while passing through a medium. It causes attenuation in the seismic amplitude.

Acceleration   Rate of change of the velocity with time.

Acoustic impedance   Seismic velocity multiplied by density.

Airgun   A seismic energy source, that releases compressed air into a fluid environment. It is the most popular marine seismic energy source.

Airgun array   A group of airguns of various sizes arranged in a geometrical pattern and fired over a very short time interval to provide a desired airgun signature.

Airgun signature   The waveform produced by an airgun array as measured at a depth of about 100 m below the array. It is intended to approximate signal waveforms produced by explosives on land.

Aliasing   It is frequency ambiguity due to signal sampling process.

Alias Filter   A filter used before sampling to remove undesired frequencies. It is also called antialias filter.

Ambient noise   Noise associated with surrounding environment.

Ammonium nitrate   A chemical compound, that is used as a fertilizer and. When mixed with diesel fuel, is used as a seismic explosive energy source.

Amplitude   The maximum swing of wave, from an average value.

Amplitude versus offset (AVO)   A method of analyzing anomalous amplitude behavior. It is performed in pre-stack data.

Analog   When applied to recorded signals, it refers to a method in which some parameter such as magnetic intensity varies continuously with the input signal.

Anomaly   In seismic usage, a deviation from uniformity in physical properties is called anomaly. It is generally synonymous with structure. Occasionally used for unexplained seismic event.

Asymmetric split spread   An arrangement of seismic sources and geophone groups along a line in which there are geophones on each side of the source but more are on one side than the other.

Autocorrelation   The cross-correlation of a wavelet with itself.

AVO gradient   The slope of a straight line fit to a plot of relative amplitude versus $\sin^2 i$, where i is the angle of incidence. CMP traces are resorted into angle traces to facilitate the plotting.

AVO intercept   The amplitude value at which a straight line fit to a plot of relative amplitude versus $\sin^2 i$ crosses the amplitude axis (where $\sin^2 i = 0$). Angle i is the angle of incidence and the intercept is equated to the vertical reflection coefficient.

Azimuth   The horizontal angle specified clockwise from true North. In 3-D survey the Azimuths are calculated for each recording line as a part of acquisition parameters.

Bandwidth   (1) The frequency band required to transmit a signal. (2) The range of frequencies over which a given device is designed to operate within specified limits.

Bin   A bin consists of cells with dimensions one-half the receiver group spacing in the inline direction and nominal line spacing in the cross-line direction (same as CMP). In data processing, sorting data in cells is called binning.

Binary   A number system with a radix of 2. It uses only the digits 0 and 1.

Binary gain   A gain control in which amplification is changed only in discrete steps by factor of 2.

Blowout   Unexpected flow of liquids or gas, due to abnormal high-pressure reservoir. The reservoir pressure exceeds the pressure of the drilling fluid.

Bow-tie effect The appearance of a buried focus on a seismic section. It is two intersecting seismic events with apparent anticline below it.

Bright spot A large increase in the amplitude due to a decrease in the acoustic impedance from the overlaying shale to the sandstone reservoir saturated with gas. As little as 5% gas saturation can cause this amplitude anomaly. It is called a direct hydrocarbon indicator for gas sands.

Bubble effect Successive expansions and contractions of the air expelled from an airgun resulting in a long duration signal wavelet.

Buried focus For zero offset and constant velocity, buried focus occurs if a reflector's center of curvature lies beneath the recording plane.

Cable feathering Cable drifting at an angle, to the marine seismic line due to crosscurrent.

Caliper log A wireline logging tool. It records the borehole diameter.

Calibration A check of equipment readings with respect to known values.

Carbonate A rock formed from calcium carbonate. Limestone and dolomite are carbonate rocks; they are potential reservoir rocks.

Casing Tubes used to keep a borehole from collapsing (caving in).

Cementing Cement is pumped to fill the space between casing and walls of the well to protect the well from collapsing.

Convolution Change in the wave shape as result of passing through a linear filter. It is a mathematical operation between two functions to obtain a desired function.

Constant angle stack In amplitude versus offset analysis, constant angle stack is applied to check the variation of the amplitude with the angle of incidence. Certain values of angles are chosen within the CMP before stack and a stack is generated.

Closure Is the vertical distance from the apex of a dome to the lowest closing contour. Areal closure is the area contained within the lowest closing contour.

Common cell gather (CCG) In 3-D seismic surveys, data are sorted in common cell gathers same as CMP in 2-D surveys.

Common depth point (CDP gather) The set of traces that have a common depth point. Each trace is from different source and receiver. It is used where there are horizontal reflectors.

Common mid-point (CMP gather) Is the set of traces that have a common mid-point, each from a different source and receiver. It is used where there are dipping reflectors.

Common shot point   Seismic data are recorded from same source and different detector stations on the ground.

Compressional waves   A p-wave or waves travel (propagate) in the same direction of motion; it propagates through the body of a medium.

Core   A rock sample cut from a borehole.

Correlation   Measures the degree of similarity between two time series, e.g. seismic traces.

Critical angle   An angle of incidence for which the refracted ray travels parallel to the surface of contact between two adjacent layers.

Cross-correlation   It is a measure of degree of similarity between two functions. In a mathematical sense, it is a cross-multiplication of sample values, addition, shift one sample and so on. Zero cross products are indication of no similarity between the functions.

Crossfeed   The induction of signal in a channel in which there is no input from a channel that has input.

Cross-line fold   Increase in multiplicity of a stack resulting from common midpoint traces present in traces recorded on adjacent detector lines.

Deconvolution   A process designed to enhance the vertical resolution of the seismic data by attenuating the undesirable signals such as short period multiples. It is also called inverse filtering.

Demultiplex   Is to rearrange the recorded digital field data in such a way that all the samples belong to every trace in the record are together in one channel. It is called trace sequential form.

Density   It is a mass per unit volume, usually measured in gram per cubic centimeter.

Diffraction   Is the phenomenon by which energy is transmitted laterally along a wave crest. When a portion of a wave is interrupted by a barrier, diffraction allows waves to propagate into the region of the barrier's geometric shadow.

Digital   A method of recording in which an analog signal is sampled at discrete times, that are usually at uniform time intervals.

Digitizer   An instrument to sample curves, seismic traces, or other data recorded in analog form.

Dim spot   Is a lack of seismic amplitude. It is caused by abnormally low reflection coefficient. Shale overlaying a porous or gas saturated reef can cause this amplitude anomaly.

Dip moveout   A process that moves data in time and space, to the correct zero-offset positions.

Dispersion   Is the variation of velocity with frequency. Dispersion distorts the shape of a wave train; peaks and troughs advance toward the beginning of the wave as it travels.

Dispersive wave   Is a wave that has changed in its shape because of dispersion. Surface waves usually suffer very large dispersion due to the near-surface velocity layering.

Down-dip   A recording geometry in which the receiver is in the direction of regional dip with respect to the source.

Downgoing wave (direct arrivals)   A seismic wave where the energy hits the detector at the top. It is widely used in the VSP terminology.

Down time   The time during which the drilling operation stopped to conduct different operations such as logging, VSP, or fishing.

Dynamic correction   Normal Move Out (NMO) is a type of dynamic correction. Dynamic correction depends on the distance from the source and the time of the seismic event. Normally, the deeper (more time), the less the dynamic correction because velocities increase with depth.

Dynamic Range   The ratio of largest signal that can be recorded to the root-mean-square (rms) value of instrument noise.

Dynamite   An explosive energy source made of nitroglycerine and other ingredients to make it stable.

Elastic   The ability for a material to return to its original shape after the distorting stress is removed.

Electrical conductivity   The ability of a material to conduct electrical current. Units are sisms per meter.

Electrical method   By which measurements at or near the earth's surface of natural or induced electric field is investigated in order to locate mineral deposits.

Enhanced oil recovery (EOR)   Techniques used for maximizing the oil production after primary recovery.

Evaporite   A sedimentary rock layer such as salt formed after water was evaporated. Other evaporites such as Gypsum and Anhydride.

Exploration   The search for commercial deposits of useful minerals , such as hydrocarbons.

Exploitation   The development of petroleum reservoirs; wells are drilled to optimally drain the reservoir and have fairly low risk. The development wells may have several producing wells in the nearby drilling or spacing units.

Extrapolation   Projection or extension of unknown value from values within known observations or interval.

Fault   Discontinuity in a rock type due to a break caused by tensional forces which cause normal faults or compressional forces which cause reverse faults.

Fault block   A geologic feature formed by faults on two or more sides.

Fermat's principle   See bow-tie effect (buried focus).

Feathering   Deviation of marine streamers towed behind a recording from a straight line caused by ocean currents.

Filters   Electrical devices or special computer software that output only a portion of the input depending on frequency, wave number, etc. of the input.

Finite Impulse Response (FIR) Filter   That component of a 24-bit recording system which transforms the single bit output stream of the Sigma-Delta Modulator into a 24-bit output and anti-alias filtering appropriate to the output sample period.

First break time   The first recorded signal from the energy source. These first breaks from reflection records are used to obtain information about the near-surface weathering layers. According to SEG polarity standard, an initial compression usually shows as a down kick.

F-K filter   A two-dimensional filter, operating in the frequency-wavenumber domains, that passes or rejects data based on apparent velocity or dip.

Flat spot   A horizontal seismic reflection caused by an interface between two fluids such as gas and water.

Fold (Structure)   An arch in a rock layer. A fold is usually formed as a result of deformation of rock layers by external forces. Folds include anticlines, synclines, and overturns, etc.

Fold (seismic)   The multiplicity of common mid-point data.

Fold taper   It is the increase in fold at the start of a seismic line and decrease in fold at the end of a seismic line. There is a similar effect in 3-D in the cross-line direction.

Formation   A distinctive lithological unit, or rock type.

Formation velocity   The speed by which, a certain type of wave travels through a particular formation.

Frequency   Repetition rate in a second of a periodic waveform. It is Measured in cycle per second or Hertz.

Frequency-wavenumber (F-K)   A domain in which the independent variables are frequency (f) and wave-number (k), yield two-dimensional Fourier transform of a seismic section, k is the reciprocal of the wavelength.

Gain   Multiplicative factor by which amplitudes are increased by seismic amplifiers, It is usually given in dB.

Gammas   A unit of magnetic intensity used to describe differences between the earth's theoretical and measured magnetic field, abbreviated as γ.

Gate   Also called window, it is a time interval where certain process or function is performed.

Geoflex   An explosive cord that has been used as a seismic energy source. It is placed near the surface from the back of a plow.

Geometric spreading   Decrease in amplitude caused by energy from a source spreading out over a larger surface area.

Geophone   A seismic detector used on land and in ocean-bottom cable systems. It is based on generation of a voltage in an electrical conductor moving in a magnetic field. Outputs are proportional to ground particle velocity.

Ghost   Type of multiple in which seismic energy travels upward and then reflects downward as occurs in the base of the weathering layer or at the surface.

GPS   An acronym for Global Positioning System, which is a Satellite-based system using orbiting satellites as the base stations.

Gravity   A method by which the subsurface geology is investigated on the basis of variations in the earth's gravitational field which are generated by a difference in rock densities. It is measured in milligal.

Ground roll   Surface-wave that travels along or near the surface of the earth. It is considered as noise on the seismic record. It is also known as Rayleigh wave.

Group interval   The distance between receiver group centers.

Harmonic distortion   The presence of harmonics of the input frequencies in the output. Harmonics are whole number multiples of signal frequency components. Distortion is measured by the ratio of the input amplitudes to total harmonic amplitude. The smaller this ratio the better.

Heterogeneous   Lateral and vertical variations of the rock properties.

Hexadecimal   A number system that has a radix of 16. It requires the use of letters A through F to supplement the digits 0 through 9.

Homogenous   Constant properties through the rock material.

Horizontal resolution   How closely two reflecting points can be situated laterally, yet can be recognized as two separate points. It is known as first Fresnel zone.

Horst   A block formed by the upthrown sides of two normal faults.

Huygen's principle   Each point on a wave front can be considered as a secondary source.

Hydrophone   A marine seismic detector that uses piezoelectric discs as the active elements. Hydrophone outputs are proportional to changes in water pressure.

Inelastic attenuation   Loss in amplitude of seismic wave caused by scattering and absorption within the medium. Shorter wavelength, hence higher frequency components, are attenuated more than longer wavelength or lower frequency components.

Inline fold   Multiplicity of a CMP stack from traces recorded along a single receiver line.

Intrabed multiple   Also called pegleg multiple. A multiple generated due to successive reflections between two different interfaces and then reflect back to the surface. Intrabed multiple has irregular travel path.

Interpolation   Determining values at location where there was no measurement. It is performed between two measured values.

Inversion (seismic)   Is to derive from the observed field data a model to describe the subsurface. Also can be used to calculate the acoustic impedance from the seismic trace.

Iteration   Procedure that repeats with improved output until some conditions are satisfied.

Least squares   An analytic function that approximates a set of data such that the sum of the squares of the distances from given points to the curve is a minimum.

Limestone   Is a sedimentary rock composed of mainly calcium carbonate. It is an important type of reservoir rock. Its matrix density is 2.7 gm/cc, and its matrix velocity is about 23,000 ft/s.

Line roll   In the swath method 3-D recording, line roll is the number of lines moved from one swath to another.

Log   A record of measurements, especially those made in a borehole (e.g, resistivity, sonic and density logs).

Love wave   A surface seismic wave characterized by horizontal motion perpendicular to the direction of propagation of the seismic wave with out vertical motion. It is near surface shear wave. Its counterpart in the p-waves is the ground roll or Rayleigh wave.

Magnetic method   A method by which the subsurface geology is investigated on the basis of variation in the earth's magnetic field. It's measured in gamma.

Magnetic susceptibility   A measure of the ability to magnetize a substance. Ferromagnetic materials Have very high susceptibilities.

Magnetic tape   A strip of plastic coated with iron oxide particles.

Matrix   A rectangular array of numbers called elements. An m x n matrix A has m rows and n columns. If m = n, it is called a square matrix.

Maximum offset   The largest distance from the source to a receiver group n 3-D recording operations.

Migration   A process used to move dipping events on seismic sections to their proper subsurface positions thereby obtaining better images of the structure and the stratigraphic picture.

Migration aperture   Is the length that should be covered in a seismic survey of a geologic feature. It is must be larger than the actual lateral extent of the feature and depends on its dip.

Minimum offset   The smallest distance from the source to a receiver group n 3-D recording operations.

Mud weight   Density of drilling mud (mass divided by volume) and expressed in pounds per gallon. The heavier the mud weight, the greater the pressure and may cause loss of circulation. Light mud weight may cause blowout if formation pressure exceeds the pressure of the mud column.

Multi-fold shooting   A seismic recording method in which fewer than half of the receiver groups in a spread are moved between shots. A number of groups are moved such that two or more traces in successive records have the same common midpoint.

Multi-offset VSP   Is a survey where a string of geophones is laid out around the well while a VSP survey is conducted. It is used to investigate the subsurface from the borehole. It is a good tool for stratigraphic implication such as sand channel mapping or delineating small faults.

Multiple   Is seismic energy that reflects more than once from the same horizon. Multiple reflections may mask out stratigraphic and structure details and is one of the undesirable signals to be attenuated.

Multiplex   A method of digital field recording in which the first sample of channel 1 is recorded, followed by the first sample of channel 2, then the first sample of channel 3 etc. until all first samples of all given channels are recorded followed by the second sample of channel 1, then the second sample of channel 2 etc.

Mute   To exclude part of the seismic data. Normally, it is applied in the early part of the traces that contains first arrivals and body waves and is called front end mute. Mute can be performed over a certain time intervals to keep ground roll, airwaves, and noise out of the stack section. This process is called surgical mute.

Navigation   The process of determining location at sea.

NMO stretch   Increase in the period (lower frequency) due to the application of normal-moveout correction to offset traces. It is noticed on the far traces within the seismic record or CMP.

Noise (seismic)   Any seismic signal but the primary reflections. This includes multiple reflection, ground roll, airwaves, source generated noise, and ambient seismic noise.

Noise test   A test or set of tests conducted in the field to analyze the noise patterns in an area to design the optimum recording parameters that will yield good signal-to-noise ratio seismic data.

Ocean-bottom cable (OBC)   A recording system used in relatively shallow water. Both geophones and hydrophones are included for each group. Some systems employ three-dimensional geophones.

Off-end spread   A method of seismic recording along a line where all geophones are on the same side of the source.

Patch   All the receivers used in recording a single 3-D record.

Permeability   Ability of the rock to transmit fluid; it is measured in millidarcies.

Permitting   The gaining of permission from landowners or appropriate authorities to operate on the land or sea where the seismic survey is to be conducted.

Petrophysical properties   Are physical aspects of the reservoir such as porosity, permeability, and fluid content.

Pilot flood   Small waterflood or enhanced oil-recovery project. It is run on small portion of a field to determine its efficiency.

Pilot sweep   A swept-frequency signal specified by starting and ending frequencies, duration of the signal and method of varying the frequency. This signal is cross-correlated with the recorded signal to produce the seismic record.

Poisson's ratio   It is an elastic constant and defined as the ratio between transverse contraction to longitudinal extension when a rod is stretched. In seismic method, it is a function of p-wave and s-wave velocities.

Poisson's ratio section   Poisson's ratio values are computed from the sum of the AVO intercept and the AVO gradient. These values are plotted versus time for each CMP to make a section.

Polarity (seismic)   The condition of the amplitude being positive or negative referred to a base line.

Porosity   Pore volume divided by bulk volume.

Preamplifiers   Constant gain amplifiers placed at the input of a seismic recording system to increase input signal levels.

Pre-plot line   The designation of desired locations of hydrophone groups for a marine seismic survey.

Primary target   The subsurface geological structure or situation that is of primary interest in the seismic survey.

Pseudo S-wave section   Pseudo S-wave values are computed from the difference between the AVO intercept and the AVO gradient. These values are plotted versus time for each CMP to make a section.

Quadratic fit   A second order approximation to get the best fit to a set of data points.

Random noise   Undesired signals without a uniform pattern; it can be attenuated by the stacking process.

Ray   A line normal to the wave front.

Ray Tracing   Determining the arrival time at detector's locations.

Rayleigh wave   See ground roll.

Reconnaissance survey   Is a survey to determine an area's main geological features. It is done to delineate an area of interest to focus on it.

Recovery factor   A percentage of what can be recovered from the reservoir fluids or gas. It varies from field to field and depends upon the geological setting of the area.

Reflection method   A technique used to investigate the subsurface by analyzing the seismic response of waves reflected from rock interfaces of different velocities and densities (acoustic impedances).

Reflectivity series   It represents reflecting interfaces and their reflection coefficients as a function of time, usually at normal incidence.

Refraction method   A technique used to map the subsurface structure by analyzing waves that enter high-velocity medium near the critical angle of incidence to the interface. It travels in the high-velocity material parallel to the refractor.

Residual statics   Trace to trace time differences on reflection events after field statics and NMO corrections have been applied. They are caused by near-surface formation and velocity irregularities. They are removed by applying refraction-based or short period statics in a surface consistent manner.

Reservoir (petroleum)   Rock containing hydrocarbon accumulation.

Root-Mean-Square velocity (VRMS)   RMS velocity approaches the stacking velocity that is obtained from velocity analysis based on the application of the normal moveout correction as the offset approaches zero. The assumption is that the velocity layering and the reflectors are parallel and that there are no changes within the layers (isotropic).

Saturation   Percentage of pore space of a certain rock filled with a particular fluid (water, oil, or gas).

SEG Y   A standard recording format adopted by the Society of Exploration Geophysicists (SEG) in 1975 for data exchange.

SEG D   A standard, multipurpose, recording format adopted by the SEG in 1980. A revised SEG D format was adopted in 1994 to accommodate changes in recording techniques.

Seismic arrays   Geometrical arrangements of receivers in a group or sources at a shotpoint that acts as filter to reject or attenuate source-generated noise.

Seismic detector   A device that detects seismic signals. Geophone on land surveys, hydrophone in marine surveys.

Seismic line   A line along which seismic data are recorded.

Seismic marker   A continuous seismic character distinguished on a seismic section.

Seismic record   A plot or display of seismic traces from a single source point; a seismogram.

Seismic resolution   Ability to separate two features close together.

Seismic section   A display of seismic data along a line. The horizontal scale is in distance units and the vertical scale is usually two-way time in seconds or sometimes in depth units.

Seismic signature   A waveform generated in a certain medium by a seismic source.

Seismic trace   "Wiggle trace" is the response of a single seismic detector to the earth's movement due to seismic energy. Each part of the wiggle trace has some meaning, either reflected or refracted energy from a layer of rock in the subsurface, or some kind of noise pattern. Excursions of the trace from a central line appear as peaks and troughs; conventionally peaks represent positive signal voltages, and the troughs negative signal voltage.

Seismic velocity   The speed by which a seismic wave travels in a particular medium. It is measured by unit distance per unit time.

Shadow zone   A portion of the subsurface from which reflections are not present because no ray path from it reached the detectors.

Shear wave (S-wave)   A seismic wave that has particle motion perpendicular to the direction of propagation. The velocity of the S-wave is approximately one-half the velocity of the P-wave.

Shot point   The location where an explosive charge is detonated. Also used for the location of any source of seismic energy.

Sigma-Delta Modulator   A single-bit analog-to-digital converter used in 24-bit seismic recording systems to sample analog signals at a very high rate.

Signal   A part of a wave that contains desired information.

Sleeve gun   An airgun that employs a movable external cylinder to release compressed air in a more uniform way than in previous models.

Slowness   Reciprocal of velocity (1/V).

Snell's law   It shows the relationships between the incidence and reflected wave.

Snell's law of reflection   States that the angle of incidence (angle between the incident ray and the normal to the interface) equals to the angle of reflection (angle between the reflected ray and the normal to the interface).

Snell's law of refraction   States that the sine of the angle of incidence divided by the velocity in the upper layer equals to the sine of the refraction angle in the second layer divided by the velocity of the second layer.

Sparse system   A large system and has about 1% nonzero values (e.g., sparse matrix).

Split spread   Arrangement of geophone groups in relation to the source point. In this case, the source point is in the middle between the geophone groups.

Spread   The receivers and source used to record a seismic record in 1-D shooting.

Spread length   The distance from source to far receiver group in 2-D shooting.

Stacking   Combining (adding) traces from different records to form a composite record. It is done in order to improve the signal-to-noise ratio and reveals subsurface geology.

Static correction (statics)   A correction applied to the seismic data to correct the irregularities of the surface elevations, near-surface weathering layer, and weathering velocities, or reference to a datum.

Stratigraphic column   A chart where the rock units are arranged from bottom (older) to top (younger) chronologically.

Streamer   A sectionalized cable consisting of active sections containing hydrophone groups and the in-water electronics, plus other sections that act as spacers and to absorb horizontal forces acting on it. It is towed behind a marine seismic vessel with other devices are built into it or attached to it that maintain streamer depth, and provide data for determining streamer shape.

Surveying The process of determining positions and elevations of seismic receiver group centers and source array centers for land seismic surveys.

Swath   A set of adjacent lines of seismic receivers along which 3-D seismic data are obtained.

Swath shooting   A method by which 3-D data are collected on land. Receiver cables are laid out in parallel lines (inline direction), and shots are positioned in a perpendicular (cross-line direction). It is also called multi-line shooting.

Symmetric split spread   An arrangement of seismic sources and geophone groups along a line in which there are an equal number geophones on each side of the source.

Synthetic seismogram   A seismic trace generated from the integration of the sonic and density logs by calculating the reflectivity series. This series is filtered with the same filter of the seismic section for better correlation. It is a man-made seismic trace and one of its applications is to transfer lithology to the seismic section.

Tesla   Unit of magnetic intensity or field strength.

Transmission coefficient   The ratio between the amplitudes of the transmitted ray and the incident ray.

Transit time   Is a measure of the sonic velocity of the rock layers and is obtained by sonic tool. Transit time is measured in microsecond per foot and varies with rock type, porosity, and fluid content.

Trap   A shape of rocks that is able to confine fluids such as oil. A trap should have a cap rock in order to prevent fluids escape. A stratigraphic trap can be formed by permeability termination.

Tomography (seismic)   The word is derived from the Greek words Tomos (section) and graphy (drawing). It is a method for obtaining models that adequately describe seismic data observations and show the effect of rock properties on the seismic wave propagation.

Tuning effect   Interference resulting from closely spaced seismic reflectors. It can cause enhancing or smearing to the individual reflection.

Unconformity   Is a buried erosional surface. It separates older rock from younger overlaying rock. Unconformities are normally good seismic markers and hydrocarbon accumulation occurs above or below the surface of the unconformity.

Up-dip   The direction opposite that of regional dip.

Upgoing wave   A seismic wave that hits the detector from the bottom after it reflects from horizons.

Uphole geophone   A geophone placed near the top of a shot hole to record energy traveling upward from the source.

Velocity pull-up   A pull-up of a reflection due to abnormal high velocity of a material such as salt.

Velocity survey   A series of measurements in a well to determine the average velocity as a function of depth. Sometimes it applies to sonic log or vertical seismic profiling (VSP).

Vertical resolution   The ability to separate two features that are very close together. Maximum vertical resolution is one-quarter of the dominant wavelength.

Vertical seismic profiling (VSP)   Is seismic survey in which a seismic signal is generated at the surface close to a well and recorded by geophone placed at various depths in the borehole.

Vertical stack   A process combining seismic records from several sources at nearly the same location without correcting static or offset differences.

Vibroseis   A land seismic energy source that inputs a vibratory signal into the ground.

Wave equation   An equation that relates the lateral (spatial) and the vertical (time) dependence of disturbances that can propagate as waves.

Wave front   A circle of equal travel time or a leading edge of a waveform.

Wave length   Velocity in unit distance per second times the period (time in seconds between two peaks or two troughs of a seismic wavelet); it's measured in unit distance. Also can be expressed as velocity divided by frequency.

Well logging   Is done by a borehole tool where a particular device can measure a variable that can be used to determine a rock property such as porosity, saturation, lithology, and formation boundaries.

Well prognosis   Prediction of geological targets before drilling.

Wildcat (well)   A well drilled in a newly explored area where hydrocarbon accumulations are not discovered commercially.

Zero-offset section   A CMP stack section where NMO corrections have the effect of moving all traces to the zero offset position.

Bibliography

Hyne, N. J. *Dictionary of Petroleum Exploration, Drilling & Production.* Tulsa, Oklahoma: PennWell Books, 1991.

Sheriff, R. E. *Encyclopedic Dictionary of Exploration Geophysics.* Tulsa, Oklahoma: SEG, 1991.

Index

Printing: Krips bv, Meppel, The Netherlands
Binding: Stürtz, Würzburg, Germany